# 4th Engineering Cosmetics and Consumer Products Conference (ECCP 2023)



Plainsboro, New Jersey, USA
18 – 20 April 2023

ISBN: 978-1-7138-8166-7

**Printed from e-media with permission by:**

Curran Associates, Inc.
57 Morehouse Lane
Red Hook, NY 12571

**Some format issues inherent in the e-media version may also appear in this print version.**



Printed with permission by Curran Associates, Inc. (2024)

For permission requests, please contact AIChE
at the address below.

AIChE
120 Wall Street, FL 23
New York, NY 10005-4020

Phone: (800) 242-4363
Fax: (203) 775-5177

www.aiche.org

**Additional copies of this publication are available from:**

Curran Associates, Inc.
57 Morehouse Lane
Red Hook, NY 12571 USA
Phone: 845-758-0400
Fax: 845-758-2633
Email: curran@proceedings.com
Web: www.proceedings.com

# TABLE OF CONTENTS

# Welcoming Remarks

As essential ingredients in personal care products, surfactants control product performance. To understand surfactant morphology, molecular dynamics simulation methods can provide insight at the molecular level. Moreover, a reliable and robust simulation method could guide the optimization of formulation designs in less time and in a more sustainable manner. As one of the widely used surfactants in shampoo and conditioner, sodium laureth sulfate (SLES), has been studied extensively. Simulation studies to date, however, can only predict surfactant morphology for a limited composition range and do not reflect the effects of solution conditions (e.g. salt dependence) on morphology. To improve the utility of these models across a wide composition range and incorporate the salt dependency, we adapted the MARTINI coarse-grained framework. Standard MARTINI force field parameters for these surfactants have thus far only been applied to predict aggregate size and micelle morphology at low surfactant concentrations without considering the impact of ionic strength. By reparameterizing the long range electrostatic interaction strengths, the modified MARTINI SLES models could successfully reproduce the salt dependent effects on SLES micelle morphology across a wide composition range. Micelle morphology and geometric properties were analyzed and are consistent with previous studies. This implementation of distance dependent electrostatic treatment in the MARTINI coarse-grained model has shown promising results on SLES molecules and can be extended to more complex formulations allowing for targeted design of new formulations.

# Forces that are converging to drive an accelerated pace of change regarding innovation anchored in the Nextgen consumer's values.

## Authors

Puma, B. - Presenter, Clorox

Next-gen consumers are the most value's-driven generational cohort today. They account for less than 20% of the population but have an outsized influence on beauty sales and word of mouth. As this group enters the workforce their values shape more than their purchases. The advent ESG investment portfolios are driving businesses to rethink how they prioritize this consumer's values in their everyday business practices so they are eligible to be a part of these funds. The strength of this consumer, combined with business prioritization of ESG practices, is also lining up with breakthrough technologies that enable entire new approaches for sourcing packaging and raw materials. This confluence of forces and technology is creating a moment in time that will drive exciting retooling of entire industries. This presentation will shed light on how these may be connected and how this may drive a new s-curve in innovation.

# Approach for a Renewed Skin Barrier.

## Authors

Mullins, M. - Presenter, Active Concepts
Topiwala, S.

The skin is our body's greatest organ, as well as the most sophisticated sensory system. Researchers have discovered that skin, like our tongue and nose, can smell and taste due to olfactory and taste receptors. This might be a starting point for designing beauty products that activate the TAS2R bitter taste receptors in human skin, contributing to beneficial cutaneous activity. In particular, the activation of bitter taste receptors through topical application may help to increase intracellular calcium levels and play a significant role in the homeostasis of the cutaneous system.

Human epidermal keratinocytes were seeded into a 96-well tissue culture plate and grown to confluency in complete media. 0.01%, 0.1%, and 1.0% concentrations of the bitter blend were added to complete media and incubated with keratinocytes. Complete media containing 1mM CaCl2 was utilized as a positive control. 2mM EGTA in the presence of 1mM CaCl2 was utilized as a negative control. After a 60-minute incubation at 37Â°C, 50Î¼L of calcium reagent solution was added. The plate was returned to 37°C for 60 minutes followed by 30 minutes at room temperature. Fluorescence measurements were taken on a Synergy H1 Microplate Reader.

The data demonstrates that the bitter blend increases intracellular calcium, an indicator of keratinocyte proliferation, differentiation, and migration, which contributes to the maintenance of skin barrier integrity and homeostasis. When these receptors are activated, the body releases calcium and protein essential to the skin barrier, and stimulates the synthesis of skin lipids. Utilizing the bitter blend delivers anti-aging effects through increasing surface metabolism, as well as the regulation of cellular senescence. Assays in Intracellular Calcium, Lipid Peroxidation, Scratch, SA-B-Gal Cellular Senescence, *in-vivo* Hydration and TEWL prove this technology can successfully activate the TAS2R bitter taste receptor in skin to effectively provide barrier strengthening, skin homeostasis, and anti-aging effects.

# Glycolipid Biosurfactants Show Potential to Positively Impact the Skin Microbiome

## Authors

**McAnerney, C. - Presenter**, Locus Fermentation Solutions
Yezhova, A., Locus Fermentation Solutions
Farmer, S., Locus Fermentation Solutions

As consumers transition to the idea of â€œclean beauty,â€ there is an increasing interest in the skin microbiome and its role in the cosmetics industry. Bacteria on the skin are an important factor in common dermatological disorders, and by removing negative agents, the overall health of the skin and microbiome would improve. Normally, harsh chemicals are employed, removing both harmful and beneficial microorganisms. The future of cosmetic additives is finding alternatives that aid the skin ecosystem.

Biosurfactants, which are known to have targeted antimicrobial activity and be mild on the skin, make excellent candidates. This study looks at combinations of primary and binary glycolipids against representative organisms. Time-kill experiments were conducted using *Staphylococcus capitis* and *Escherichia coli* for gram positive and negative bacteria, respectively. Within 20 seconds, the population of *S. capitis* was reduced by nearly 50%, and after 10 minutes it was further decreased to 5%. Thus, the blend of glycolipids showed a strong ability to kill Gram-positive bacteria. Gram-negative bacteria, which are commonly more resistant to stressors, showed a greater vulnerability to these glycolipids. When tested against *E. coli*, the biosurfactant combination diminished the population by 95% within 20 seconds and after 10 minutes the viable cell count was 0%. These results show strong effects against potentially antagonistic bacteria located within the skin microbiome. Based on the efficacious antimicrobial effect that was observed, the application of these glycolipids in the cosmetics industry would have a clean influence that aids the microbiome.

# Leveraging Sustainable Ingredient Design into a Corporate ESG Program

## Authors

Wesson, A. - Presenter, Inolex

At Inolex, we design sustainable ingredients for health, beauty, and wellness guided by the Principles of Green Chemistry and life cycle thinking. Our imaginative approaches are the building blocks for exceptional products that care for people and our planet. Fittingly, our products also became the building blocks for a formalized corporate sustainability program. As chemists and engineers, we approach sustainability differently than business-minded stakeholders. In this talk I will explore the interplay of the sustainable design work with the development of a corporate sustainability program and how the two mindsets, engineering and business, can be leveraged into a successful and powerful sustainability engine.

# Cellulose Esters As Sustainable Alternatives to Microplastics in Personal Care Applications

## Authors

Bhattacharya, D. - Presenter, Eastman Chemical

In recent years, there has been a strong demand from consumers that the products they use meet clean beauty standards. Increasing consumer awareness, along with regulatory pressure has driven raw material suppliers and brand owners to find innovative solutions for sustainability. Many cosmetic products contain microplastics that persist in the environment indefinitely and are being inadvertently consumed by wildlife and humans. Eastman has been addressing these challenges by leveraging over 100 years of material innovation to provide solutions to consumer brands for color cosmetics, skin/sun care, and hair styling applications. A key innovation focus for Eastman is to develop biobased and biodegradable alternatives to the unmet need of cosmetic formulators as a replacement of microplastics by utilizing our expertise in cellulosic chemistry.

# Potential of Recombinant Proteins Made By Biotechnology in Cosmetics

## Authors

O'Neill, J. - Presenter, Modern Meadow
Hansenne-Cervantes, I. - Presenter, Modern Meadow

Collagen and its derivative proteins have been widely used as a major compound for skin care products as a natural ingredient and moisturizer. Most commercially available collagens are animal-derived collagen type I and other forms of collagen, such as type III collagen, are far less prevalent in animals, making extraction and purification extremely difficult and expensive. We report the production of a 50 kDa protein produced in yeast that is 100% identical to the N-terminus of the human type III collagen. This recombinant protein has a larger molecular weight than most incumbent recombinant collagen proteins available for personal care applications. We have managed the industrialization of both the fermentation and purification processes to produce a final recombinant protein product. This final protein product was shown to be safe for application to human skin and compatible with common formulations. This recombinant collagen type III protein was also shown to uniquely stimulate collagen type III production and secretion by primary human dermal fibroblasts. The unique combination of biostimulation, compatibility with beauty product formulations and demonstrated commercial production, make this novel recombinant type III collagen a good candidate for broad application in dermatology and in cosmetics.

Two clinical studies have been performed to evaluate the interest of the Bio-Coll@gen in anti-ageing with instrumental quantification of the collagen content, the barrier function and the skin density. In addition the clinical experts have scored the improvement of different parameters and the satisfaction of the consumers has been investigated with a questionnaire at different periods of time.

# Ethyl Macadamiate as a Nature Derived Silicone Alternative in Personal Care

## Authors

Quinn, T. - Presenter, Cargill

Silicone-free claims continue to grow year over year as consumers seek more naturally derived, sustainable ingredients. Brands are continually looking for silicone alternatives to increase naturality without sacrificing the silky, smooth skin feel provided by silicones. Ethyl macadamiate (derived from macadamia oil) is a light ester that offers consumers a similar product aesthetic and sensory experience to low viscosity silicones, but has the added benefit of being 99.95% nature derived. The macadamia tree is promoted as an alternative crop to small farmers in developing regions of the world since it has a relatively low input requirement, and it can continue to yield seeds for decades. It can consequently provide crops for a variety of industries and allow growers to maximize the benefits of the land while having minimal impact compared to other crops. Ethyl macadamiate is created through the transesterification of macadamia oil with ethanol, yielding a unique blend of fatty acid ethyl esters. The end product is a versatile ester, with a similar skin feel to low viscosity silicones, that increases skin hydration and allows formulators to reduce or replace silicones entirely.

# Simultaneous Analysis of Multiple Colloidal Components in Complex Cosmetics Using Total Holographic Characterization

## Authors

Philips, L. - Presenter

Cosmetics are complex combinations of multiple emulsions and particle types. Total Holographic Characterization easily distinguishes these different components and determines size, composition, morphology and concentration of each component simultaneously, even when the different emulsion droplets and particles are the same size. Total Holographic Characterization measures particle holograms to distinguish multiple components in complex colloidal mixtures without special sample preparation or separation in a single measurement. In the work presented, Total Holographic Characterization is used to characterize heterogeneous suspensions containing different types of emulsions and particles in the size range from 500 nm to 10 Î¼m, including complex mixtures found in typical cosmetic products. Traditional light scattering techniques can measure the distribution of particle sizes, but are unable to distinguish different species of sub-visible particles that are the same size when they are present in the same sample. Index of refraction is different for different emulsions and particles so that each component can be distinguished and characterized.

# Beauty Devices and Connected Beauty, Success vs. Overload

## Authors

Akridge, R. - Presenter, Opulus Beauty
Over the past 30 years, innovators have created various personal care devices to make our lives easier and give us superior results, from toothbrushes to facial cleansing devices. We will discuss what it takes to invent a successful device, the biggest challenges to making your invention successful, and how far you should go to connect to the digital world. Iâ€™ll present relevant examples that I have had a personal connection with to hopefully guide you and inspire you to make a difference in our world of beauty.

How AI is Changing Consumer-Centric Product Design
Lai Hing Vair
4th Engineering Cosmetics and Consumer Products Conference
April 20, 2023


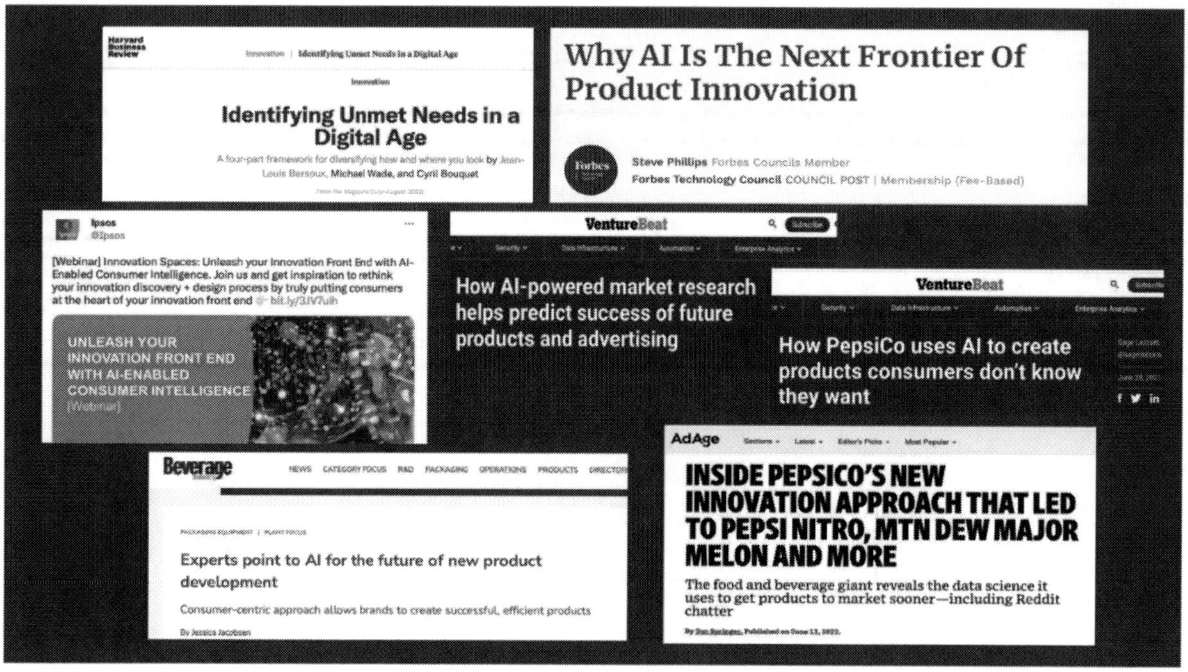

Harvard Business Review
Innovation | Identifying Unmet Needs in a Digital Age
Identifying Unmet Needs in a Digital Age
A four-part framework for diversifying how and where you look by Jean-Louis Bersoux, Michael Wade, and Cyril Bouquet
Why AI Is The Next Frontier Of Product Innovation
Steve Phillips Forbes Councils Member
Forbes Technology Council COUNCIL POST | Membership (Fee-Based)
Ipsos
@Ipsos
[Webinar] Innovation Spaces: Unleash your Innovation Front End with AI-Enabled Consumer Intelligence. Join us and get inspiration to rethink your innovation discovery + design process by truly putting consumers at the heart of your innovation front end bit.ly/3JV7uih
UNLEASH YOUR INNOVATION FRONT END WITH AI-ENABLED CONSUMER INTELLIGENCE
[Webinar]
VentureBeat
How AI-powered market research helps predict success of future products and advertising
VentureBeat
How PepsiCo uses AI to create products consumers don't know they want
Beverage
NEWS CATEGORY FOCUS R&D PACKAGING OPERATIONS PRODUCTS
Experts point to AI for the future of new product development
Consumer-centric approach allows brands to create successful, efficient products
By Jessica Jacobsen
AdAge
Sections Latest Editor's Picks Most Popular
INSIDE PEPSICO'S NEW INNOVATION APPROACH THAT LED TO PEPSI NITRO, MTN DEW MAJOR MELON AND MORE
The food and beverage giant reveals the data science it uses to get products to market sooner—including Reddit chatter

## How is AI Changing Consumer-Centric Product Design?

- AI is changing consumer-centric product design by allowing companies to better understand consumer needs and preferences, customize products to individual customers, personalize marketing campaigns, and automate product development processes.
- AI-powered solutions can help companies identify user trends and patterns, recommend products, and collect feedback to continuously improve products.
- AI can also be used to develop artificial intelligence-based products and services that can be customized to meet customer demands.

- OpenAI's Davinci 003 model

## Value of AI in Consumer-Centric Product Design

FASTER…
DECISION MAKING,
DEVELOPMENT,
SPEED TO MARKET

INCREASED
EFFICIENCY
(less cost & effort)

DEEPER INSIGHTS
→ BREAKTHROUGH
INNOVATION

ENGAGE
CONSUMERS IN NEW
WAYS

BUILD TRUST WITH
OUR CONSUMERS

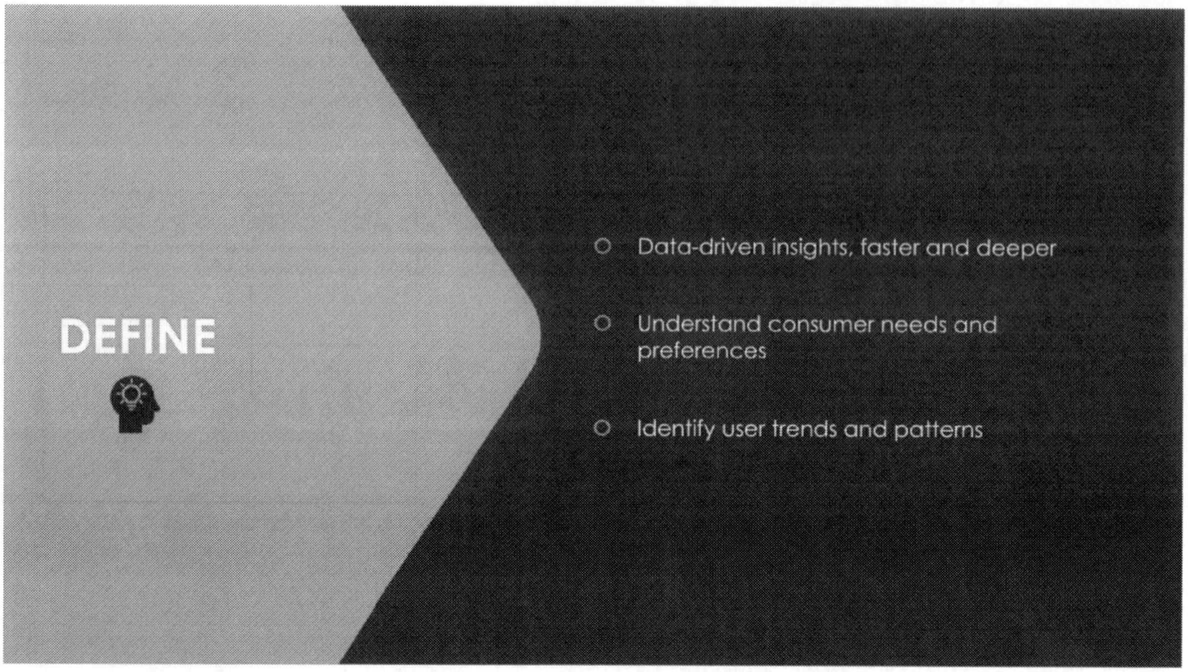
DEFINE
Data-driven insights, faster and deeper
Understand consumer needs and preferences
Identify user trends and patterns

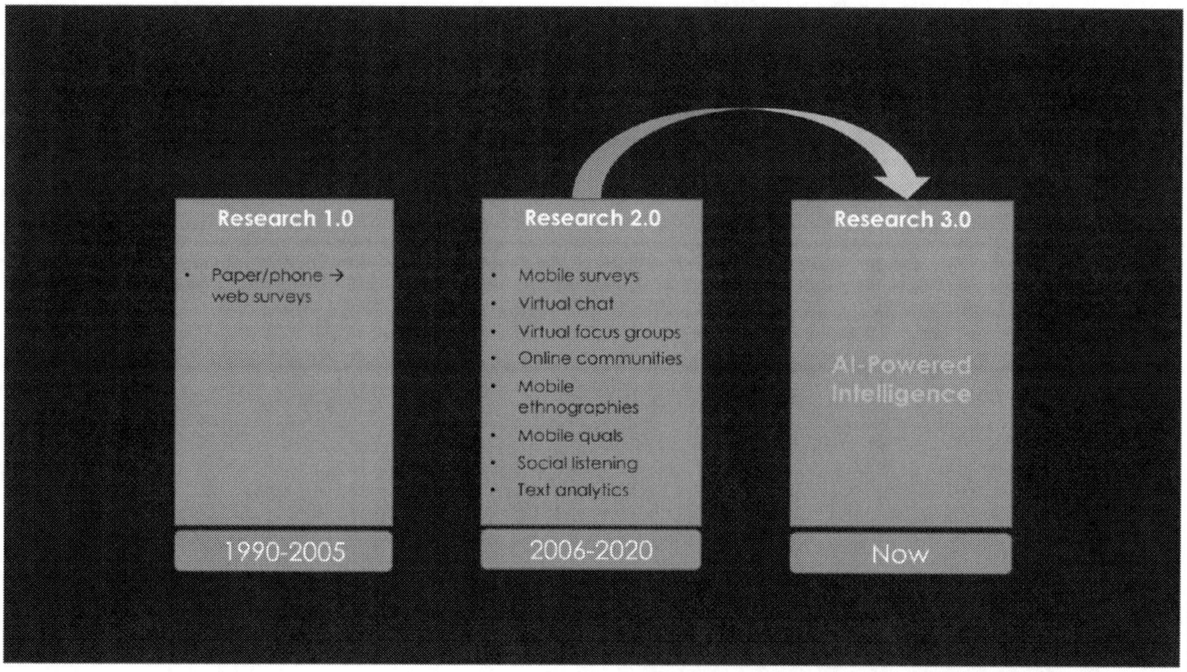
Research 1.0
Paper/phone → web surveys
1990-2005
Research 2.0
Mobile surveys
Virtual chat
Virtual focus groups
Online communities
Mobile ethnographies
Mobile quals
Social listening
Text analytics
2006-2020
Research 3.0
AI-Powered Intelligence
Now

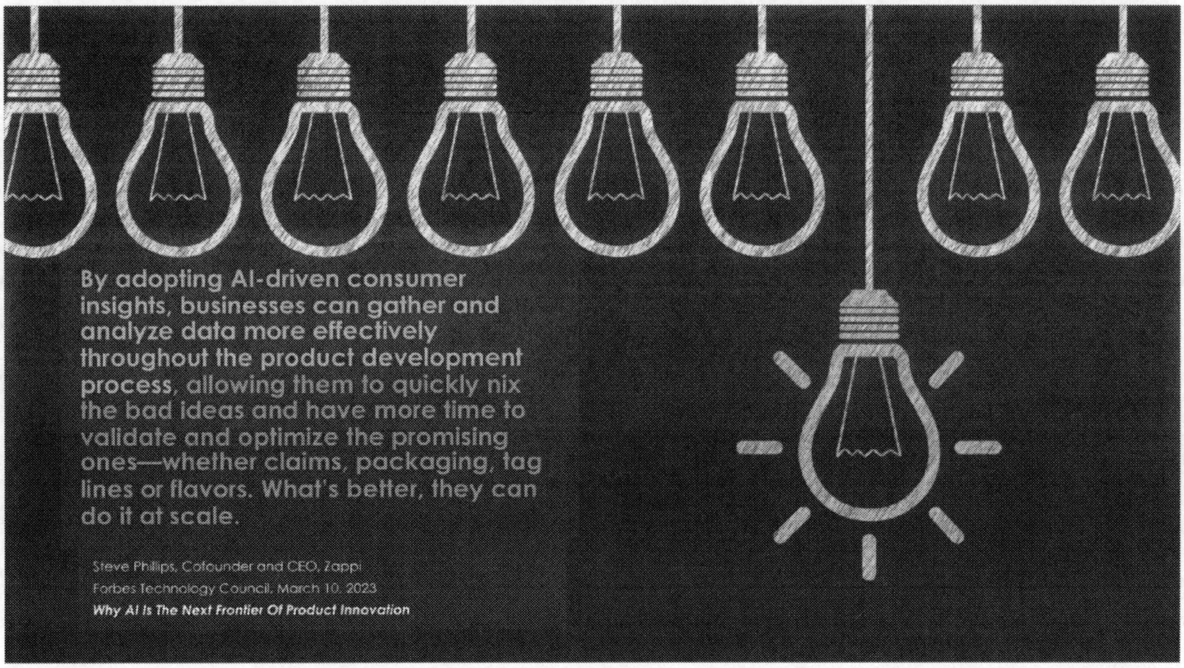
By adopting AI-driven consumer insights, businesses can gather and analyze data more effectively throughout the product development process, allowing them to quickly nix the bad ideas and have more time to validate and optimize the promising ones—whether claims, packaging, tag lines or flavors. What's better, they can do it at scale.
Steve Phillips, Cofounder and CEO, Zappi
Forbes Technology Council, March 10, 2023
*Why AI Is The Next Frontier Of Product Innovation*

AI-Powered Enterprise Intelligence
Lucy®, Your AI-Powered Enterprise Search Answer Engine-
Lucy
Lucy gets you the right answer, at the right time, no matter where it sits in your organization.
Lucy® Launches LucyGPT, An AI-Powered Subject Matter Expert that Liberates Corporate Knowledge
KUMARAN PEPSI @KumaranPepsi · Dec 11, 2021
Our #PepsiCo digital AI assistant, Ada, is using data to provide faster consumer insights than ever.
ada PEPSICO
Meet Ada: PepsiCo's Intelligent Assistant
Using AI to provide human insights
Stravito
A better way to find your insights
Stravito is a search and information discovery tool for all your market research reports and insights.
AI-powered search
For all your unstructured data
Quick to implement and easy to use

# From Social Listening to Social Intelligence

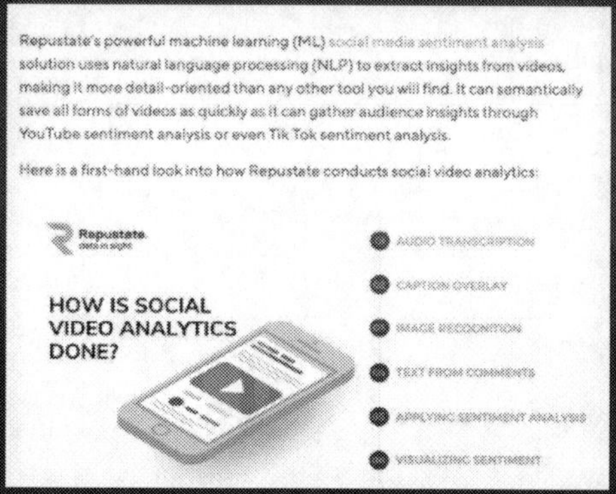


# AI-Powered Trend Prediction

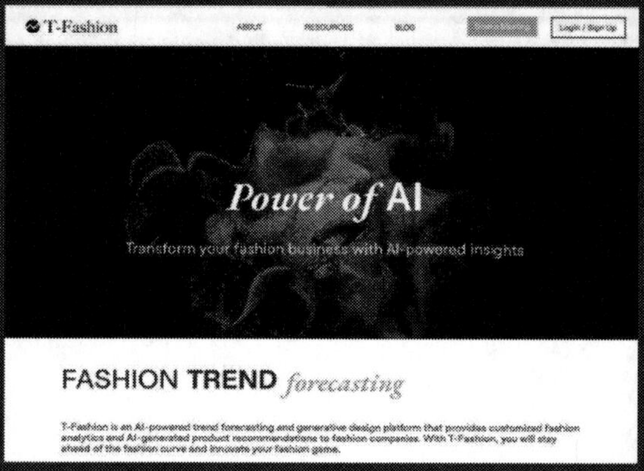

# AI-Powered Sensory Intelligence

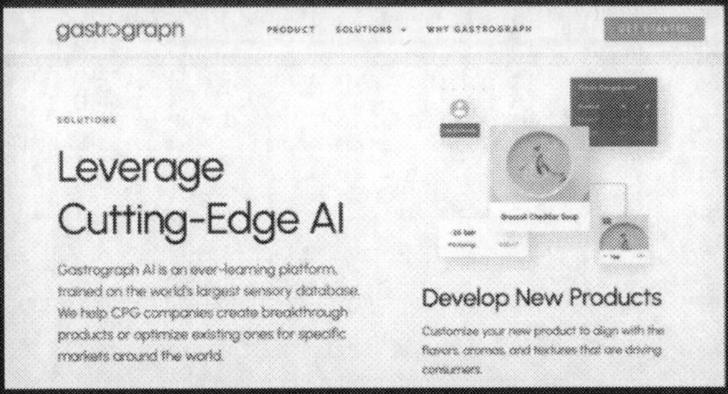


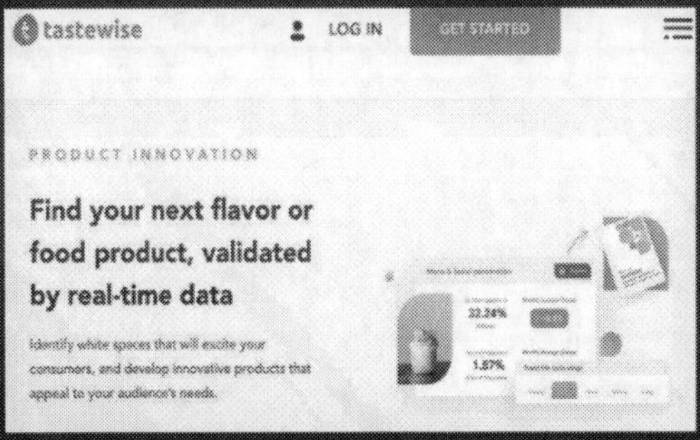


# Agile Iterative Product Concept Testing

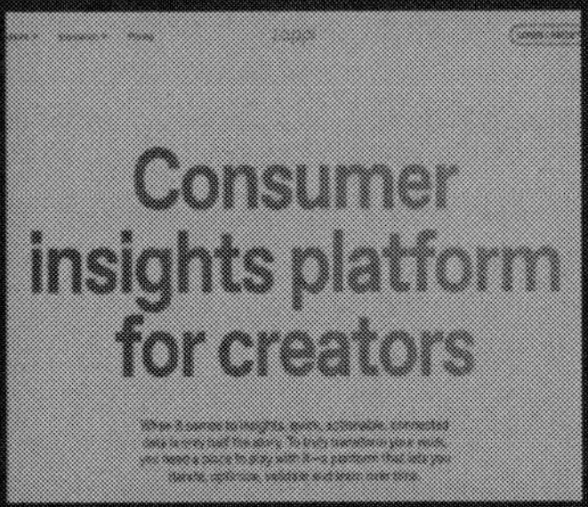


$20,000 and 4-6 weeks → $2,000 and 4-6 hours

images/designs

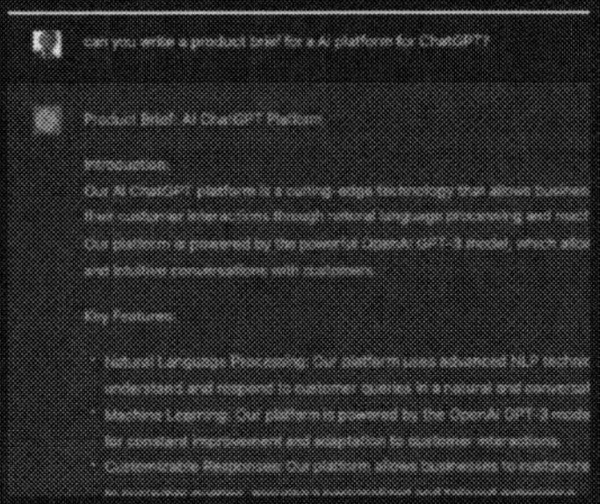

product concepts

## DESIGN & DEVELOP

- Create and iterate on new designs
- Guide and simulate with predictive models
- Optimize designs

## AI-Generated Designs

*"I told the AI to design a minimalistic sports car inspired by a MacBook and a Magic Mouse built out of aluminum and glass, while also specifying it to create something in the style of Apple's former design head, Jony Ive"* says Mauriello.

https://designwanted.com/ai-generated-apple-car-futuristic-stunning/

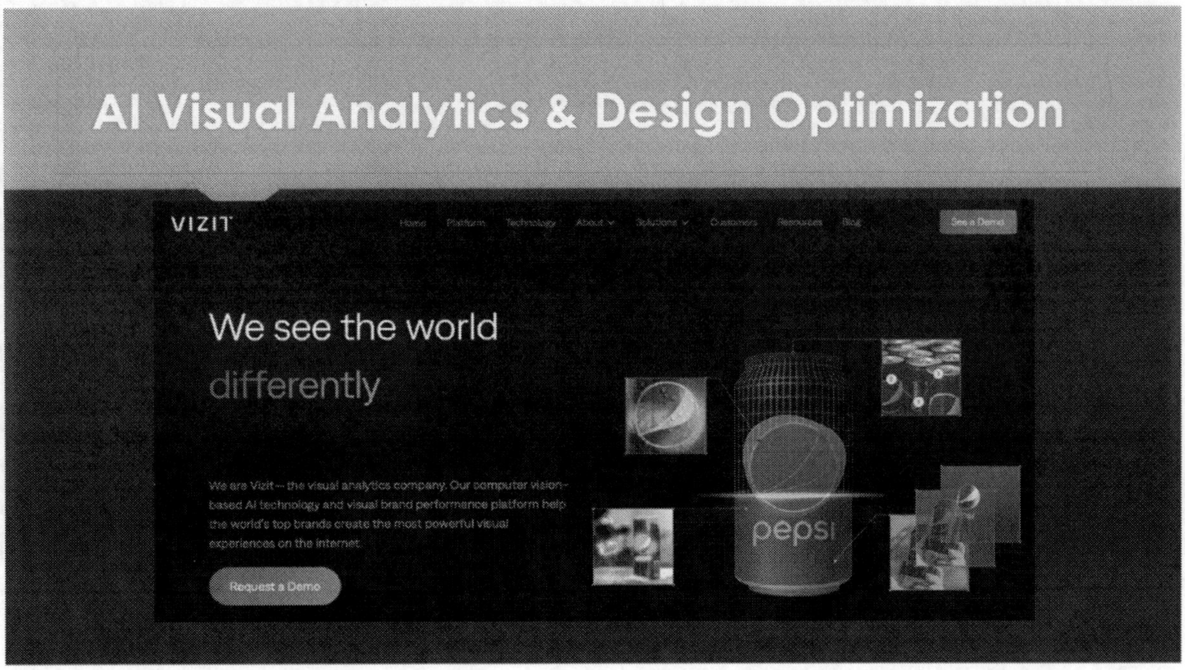
AI Visual Analytics & Design Optimization
VIZIT
We see the world
differently
We are Vizit — the visual analytics company. Our computer vision-based AI technology and visual brand performance platform help the world's top brands create the most powerful visual experiences on the internet.
Request a Demo
pepsi

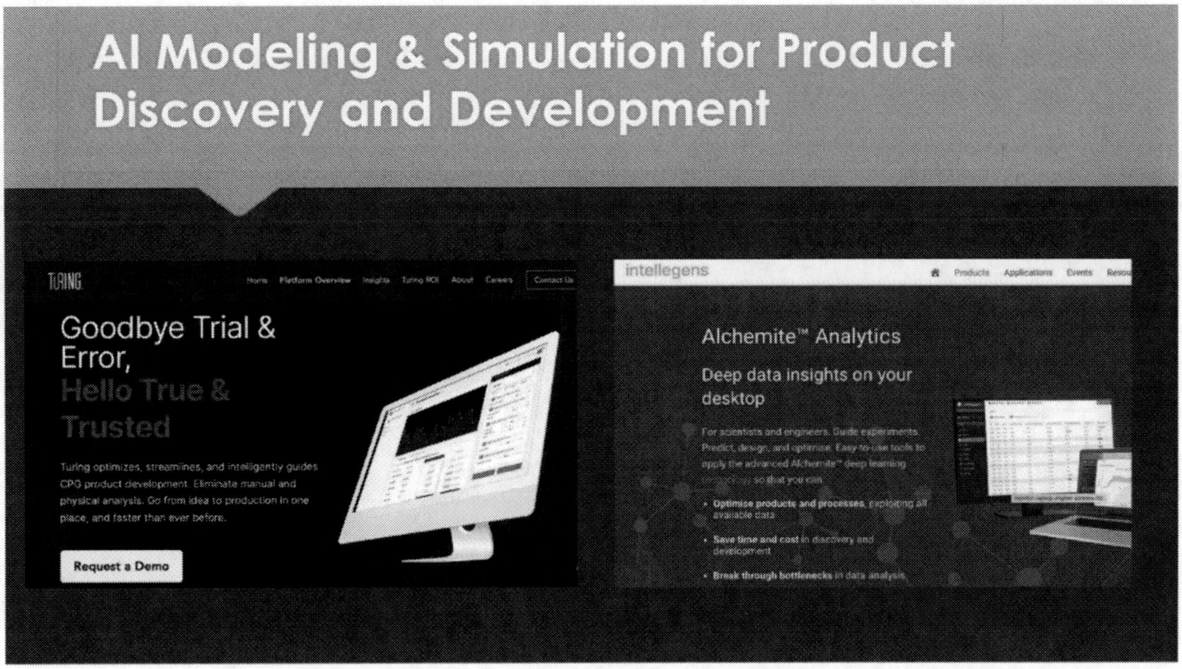
AI Modeling & Simulation for Product Discovery and Development
Goodbye Trial & Error,
Hello True & Trusted
Turing optimizes, streamlines, and intelligently guides CPG product development. Eliminate manual and physical analysis. Go from idea to production in one place, and faster than ever before.
Request a Demo
intellegens
Products
Applications
Events
Alchemite™ Analytics
Deep data insights on your desktop
Optimise products and processes, exploiting all available data
Save time and cost in discovery and development
Break through bottlenecks in data analysis.

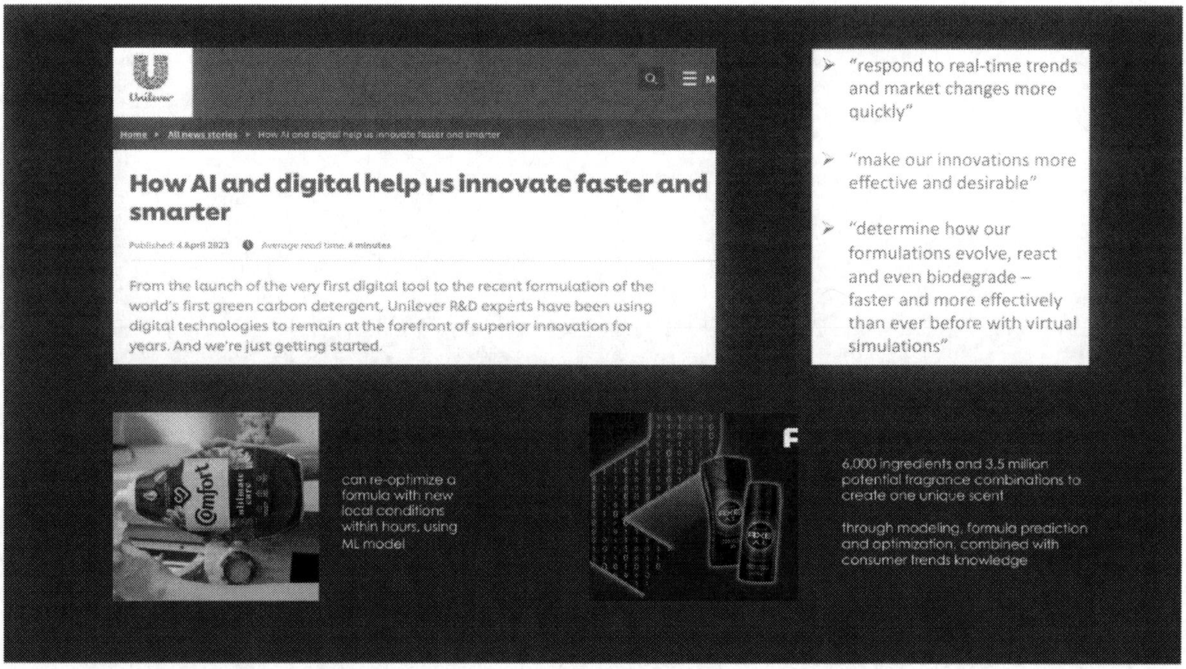
Unilever
Home > All news stories > How AI and digital help us innovate faster and smarter
How AI and digital help us innovate faster and smarter
Published: 4 April 2023
Average read time: 4 minutes
From the launch of the very first digital tool to the recent formulation of the world's first green carbon detergent, Unilever R&D experts have been using digital technologies to remain at the forefront of superior innovation for years. And we're just getting started.
"respond to real-time trends and market changes more quickly"
"make our innovations more effective and desirable"
"determine how our formulations evolve, react and even biodegrade – faster and more effectively than ever before with virtual simulations"
Comfort
can re-optimize a formula with new local conditions within hours, using ML model
6,000 ingredients and 3.5 million potential fragrance combinations to create one unique scent
through modeling, formula prediction and optimization, combined with consumer trends knowledge

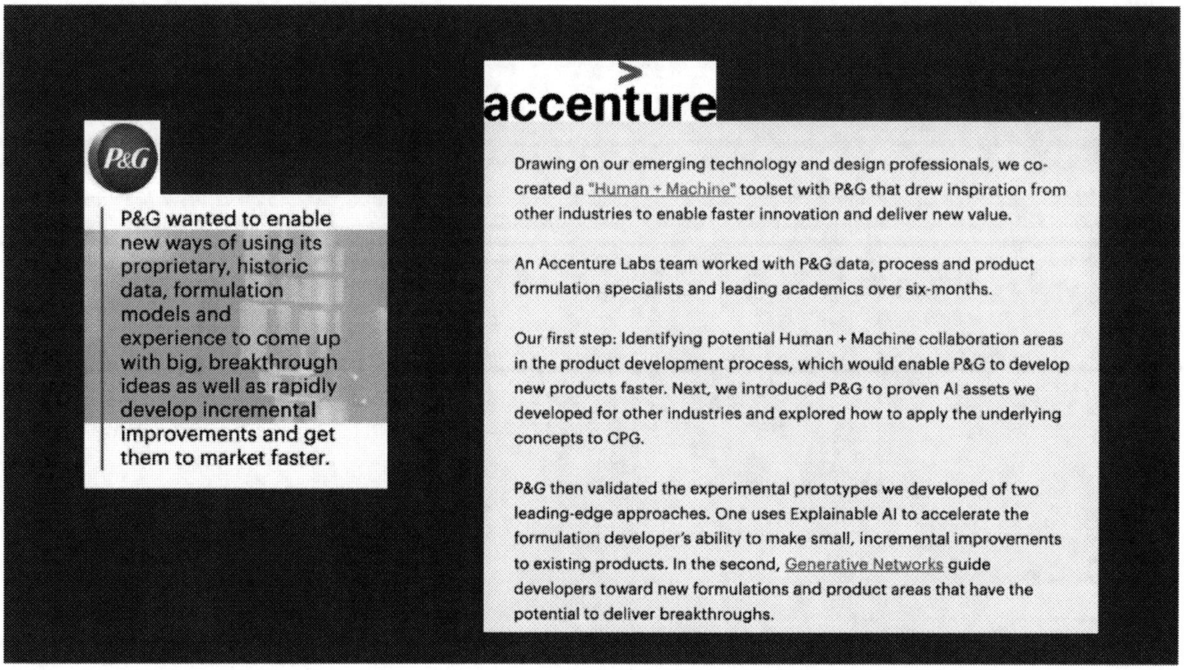
P&G
P&G wanted to enable new ways of using its proprietary, historic data, formulation models and experience to come up with big, breakthrough ideas as well as rapidly develop incremental improvements and get them to market faster.
accenture
Drawing on our emerging technology and design professionals, we co-created a "Human + Machine" toolset with P&G that drew inspiration from other industries to enable faster innovation and deliver new value.
An Accenture Labs team worked with P&G data, process and product formulation specialists and leading academics over six-months.
Our first step: Identifying potential Human + Machine collaboration areas in the product development process, which would enable P&G to develop new products faster. Next, we introduced P&G to proven AI assets we developed for other industries and explored how to apply the underlying concepts to CPG.
P&G then validated the experimental prototypes we developed of two leading-edge approaches. One uses Explainable AI to accelerate the formulation developer's ability to make small, incremental improvements to existing products. In the second, Generative Networks guide developers toward new formulations and product areas that have the potential to deliver breakthroughs.

## DELIVER

- Consistently deliver quality products to consumers
- Customize and recommend products to individual customers
- Collect feedback to continuously improve products

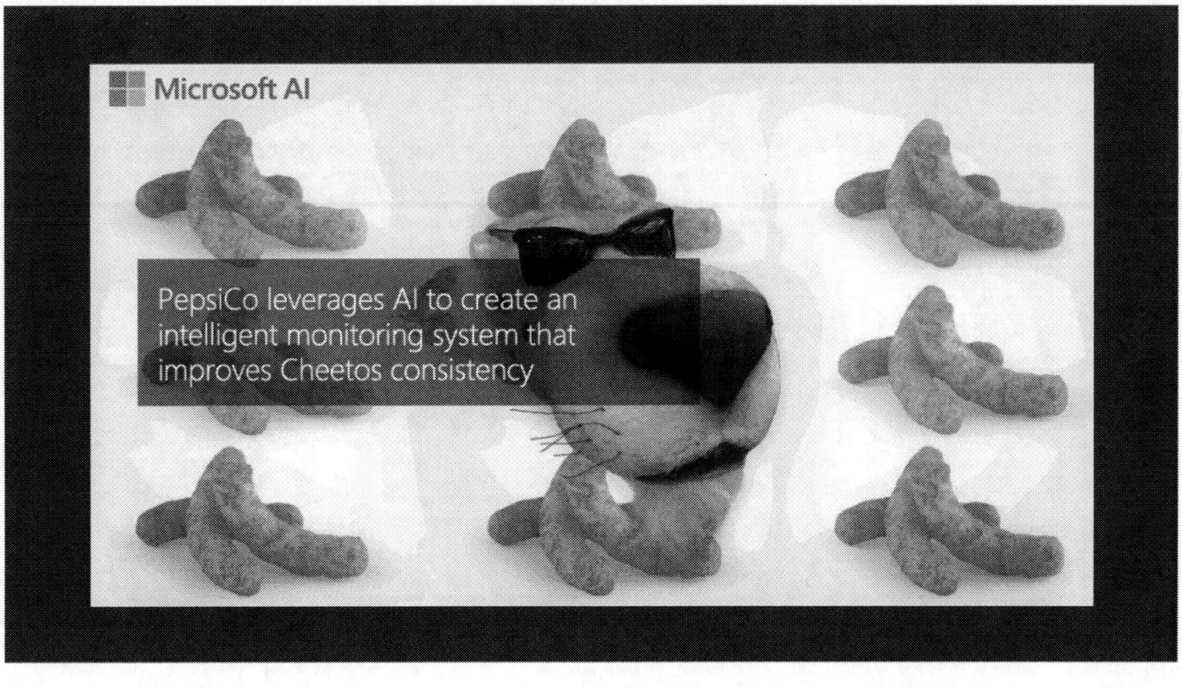

# AI to Analyze, Customize, and Optimize

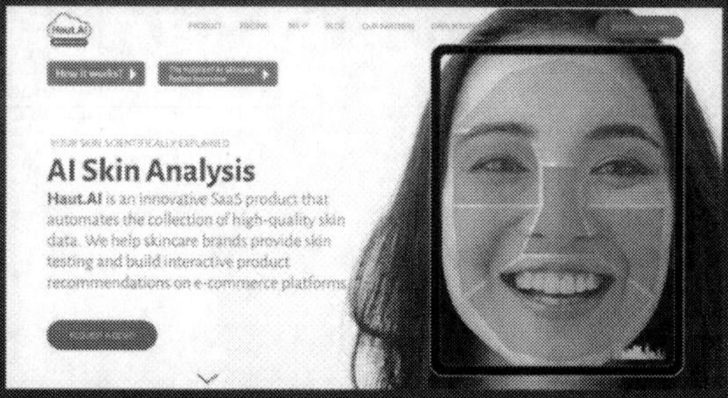

# Considerations

- Ethical and responsible AI
  - Risks, e.g. privacy, security, intellectual property, copyright
  - Algorithm bias

*"What all of us have to do is to make sure we are using AI in a way that is for the benefit of humanity, not to the detriment of humanity." – Tim Cook, Apple*

- Future of work; hybrid human-artificial intelligence

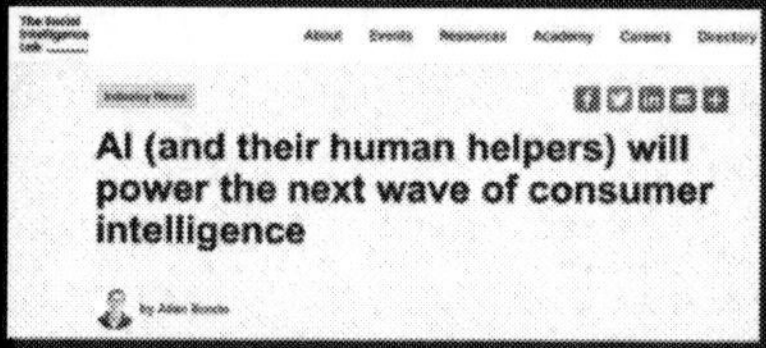

"The future of AI is both exciting and uncertain, but with a little humor and creativity, we can face it head on. Remember, it's not the machines we should fear, but our own lack of imagination."
THE LAST STAND OF THE HUMANS
A HUMOROUS GUIDE TO FIGHTING BACK AGAINST AI
Shiva S. Mohanty

# Wearable Devices and the Human Skin: Future Opportunities or a Challenge for Personal Care?

## Authors

Menon, G. - Presenter, California Academy of Sciences

Research on Wearable electronic devices is a rapidly growing field of Biomedical engineering. These devices have diverse applications â€“ monitoring health, delivering drugs and for providing a â€œskinâ€ which imparts haptic sensation for robots etc. Bio-mimetic second skins that draw inspiration from human skin has progressed from Robotic to human applications with the aim of imparting and even surpassing some of the sensory functions of skin.

Biomedical engineers have exploited an array of piezoelectric polymers, nano-technology for generating ultrasound, magnetic and electric field â€“ and even energy-independent e-skins for continuous operational efficacy. While I am eminently unqualified to discuss the technology behind these e- skins, I will discuss potential impact of this technology on the human skin, based on our current knowledge of skin biology, skin barrier, and the skinâ€™s dynamic interactions with the environment. The outermost layer of skin- the Stratum corneum (SC), recognized as a â€œsmart materialâ€ In itself is our interface with the environment, receiving signals from outside and transmitting (mechano-transduction) to the viable layers of epidermis to elicit appropriate adaptive responses. The e-skins are in fact inserted between the SC and the external world, purportedly for improving the skin functions. Possibly some of the devices could prove to be very beneficial, and may even have anti-aging effects. For the personal care industry, they could start a new trend (or a fad), with as yet unknown benefits. However, we may be trying to outsmart the smart material (SC) which has evolved over several millions of years. The possibility of some undesirable side effects on skin functions cannot be ignored at this point. This presentation will examine some of these unknown unknowns (borrowing from Rumsfeld) within the limited time, and stress the need for enough clinical studies.

# Multi-Scale Simulation for Developing and Optimizing Cosmetics Formulations

## Authors

Longwell, G. - Presenter, Siemens Digital Industries Software
Aglave, R., Siemens PLM Software
Sweere, T. J. M.

Digital transformation - As companies implement better â€œdigitalâ€ systems to store and share experimental data across their organization, this data becomes more easily accessible to computational chemists to create and validate molecular models. Using the laws of statistical mechanics, these generated molecular models can be used to predict properties of complex formulations and unexplored cosmetics products.

We will showcase how to simulate the macro-scale product behavior of complex formulations using physics-based calculations on a molecular level. To do this we cover all aspects of multi-scale simulations in chemistry, crossing relevant length and time scales giving scientists the opportunity to model properties related to materials and processes. This enhances product development and helps material scientists to design novel formulations with the desired properties. We will highlight a few examples that include:

One example comprises stability calculations of oil-in-water emulsions to enhance the shelf life of cosmetics formulations. By designing a digital model for the complex formulation, and automatically assigning the proper set of molecular interaction parameters, it is possible to investigate the behavior of the emulsifier at the interface on a molecular level. From there we can calculate and minimize the interfacial tension of stabilized oil-water interfaces by optimizing the ingredient proportions and/or emulsifier structure.

Another example, the viscosity of a consumer product highly influences its processability. For example, for ointments and products in tubes, only a narrow and consistent range of viscosities is acceptable for comfortable use. Natural polysaccharides are additives to tune the viscosity, however, variations in temperatures and/or ingredients can affect the end result. We will show how we obtain rheological properties using accurate and fast coarse-grained simulations. The obtained rheological data is useful for ingredient optimization and can also serve as input for Computational Fluid Dynamics models for simulation of larger length and time scales.

# Towards Accurate Coarse-Grained Simulation of Sodium Laureth Sulfate(SLES) Morphology

## Authors

Liu, H. - Presenter
Sanders, J., Schrodinger
Shelley, J., Schrodinger
Browning, A., Schrodinger, Inc.
Misra, M., Cornell University
Fallah, F., Schrödinger, Inc.
Giesen, D. J.
Halls, M. D., Schrodinger, Inc.

As essential ingredients in personal care products, surfactants control product performance. To understand surfactant morphology, molecular dynamics simulation methods can provide insight at the molecular level. Moreover, a reliable and robust simulation method could guide the optimization of formulation designs in less time and in a more sustainable manner. As one of the widely used surfactants in shampoo and conditioner, sodium laureth sulfate (SLES), has been studied extensively. Simulation studies to date, however, can only predict surfactant morphology for a limited composition range and do not reflect the effects of solution conditions (e.g. salt dependence) on morphology. To improve the utility of these models across a wide composition range and incorporate the salt dependency, we adapted the MARTINI coarse-grained framework. Standard MARTINI force field parameters for these surfactants have thus far only been applied to predict aggregate size and micelle morphology at low surfactant concentrations without considering the impact of ionic strength. By reparameterizing the long range electrostatic interaction strengths, the modified MARTINI SLES models could successfully reproduce the salt dependent effects on SLES micelle morphology across a wide composition range. Micelle morphology and geometric properties were analyzed and are consistent with previous studies. This implementation of distance dependent electrostatic treatment in the MARTINI coarse-grained model has shown promising results on SLES molecules and can be extended to more complex formulations allowing for targeted design of new formulations.

# In Situ Characterization of the Structural Changes Induced By Acidity Fluctuations in Hydrated Collagen Hydrogels

## Authors

Bronner, O. - Presenter
Vidavsky, N.
Nativ-Roth, E., Ilse Katz Institute For Nanoscale Science & Technology, Ben-Gurion University

Collagen is the main component of the extracellular matrix (ECM). In the ECM, The collagen forms *in vivo* a complex network and a hydrated gel-like environment, which can be mimicked *in vitro* using a collagen hydrogel. Water plays a crucial role in the self-assembly of collagen fibers by forming hydrogen bonds between residues of the polypeptide chain. The collagen structure can be affected by many factors and is directly related to its function. As a significant part of the ECM, collagen may experience changes in the chemical environment in pathological conditions. For example, in proximity to solid tumors, the ECM acidity increases, while in chronic wounds, it decreases. To mimic ECM initially under healthy physiological conditions that then changed to pathological, we characterize structural changes that collagen hydrogel undergoes when its pH changes from neutral to acidic or basic while maintaining physiological temperature and ionic strength. We investigate the hydrogel structure as closely as possible to its native hydrated state to avoid possible conformational changes during drying. Using Fourier Transform InfraRed Spectroscopy (FTIR) without drying, we show that the collagen secondary structure does not change when the pH changes from neutral to acidic or basic. Cryo Scanning Electron Microscopy (Cryo-SEM) of high-pressure frozen and freeze-fractured collagen gel reveals its submicrometer structure. When the neutral collagen pH decreased, its fibrils were shorter and lacked spatial orientation. The fibrils were helical, longer, and spatially oriented when the pH increased. When returning from acidic to neutral collagen pH, the original structure of the collagen fibers was restored. These findings may contribute to a deeper understanding of the mechanistic basis of pathological conditions in which the collagen structure plays a role in disease progression. In addition, insights from this study may help optimize collagen-based scaffolds for specific tissue engineering applications.

Spheryx
solutions for suspensions

# Simultaneous Analysis of Multiple Colloidal Components in Complex Cosmetics using Total Holographic Characterization

Fook Chiong Cheong, Juliana Lumer, Mary Ann Odete, and Laura A Philips

April 2023

## Total Holographic Characterization of Complex Mixtures

- Total Holographic Characterization® (THC)
- How it works and validation
- THC of Complex Mixtures:
  - *Size and refractive index distributions*
- Applications to Cosmetics

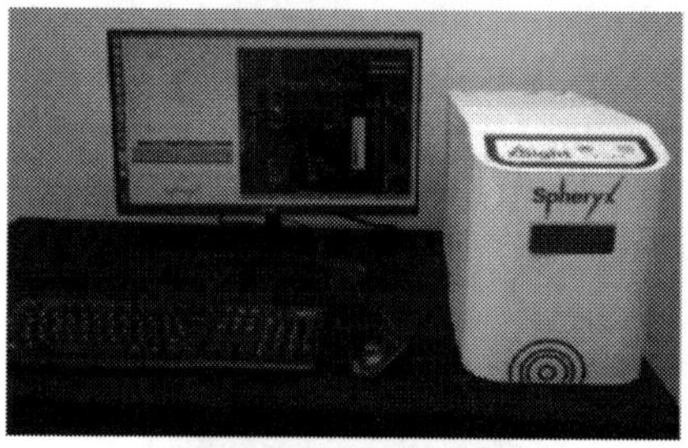

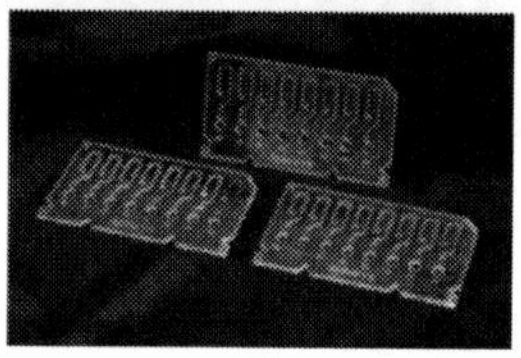

Spheryx

2

## Total Holographic Characterization: Measurement

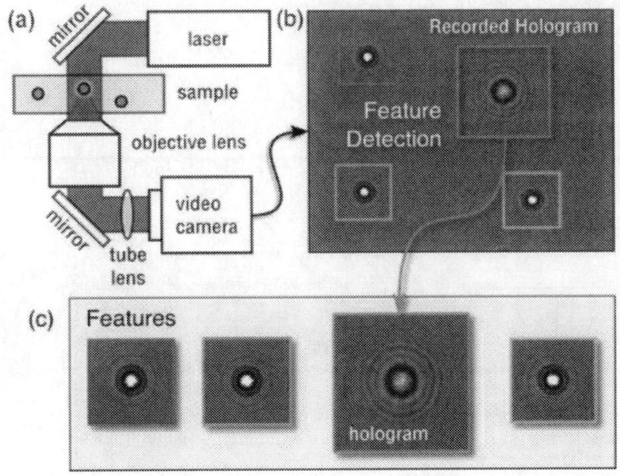


a) Measure: Particles flow through laser beam in a microfluidic channel
b) Record: Microscope records holograms
c) Detect: Each hologram detected One particle ⇒ One hologram

## Total Holographic Characterization: Analysis

**Spheryx's software analyzes holograms using Lorenz-Mie light scattering theory**

**Each fit yields**

- **diameter: $d_p$**
- **refractive index: $n_p$**
- **3D position**
- **Symmetry**

**One particle ⇒ One point**
**Concentration = # of particles/vol**

Unique ability to identify both composition and size

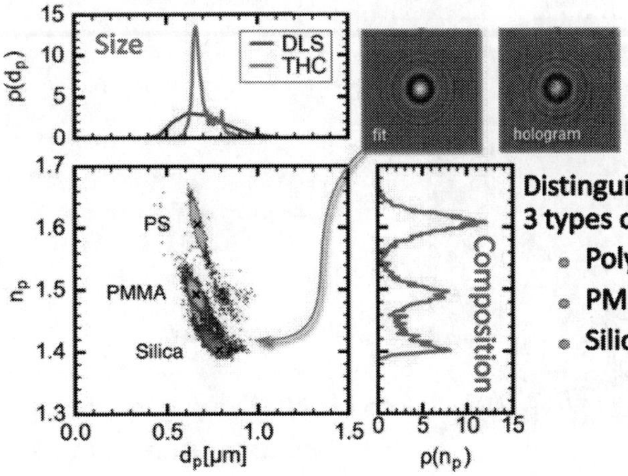


**Distinguishes 3 types of beads:**

- **Polystyrene**
- **PMMA**
- **Silica**

# Robust, Reproducible Results

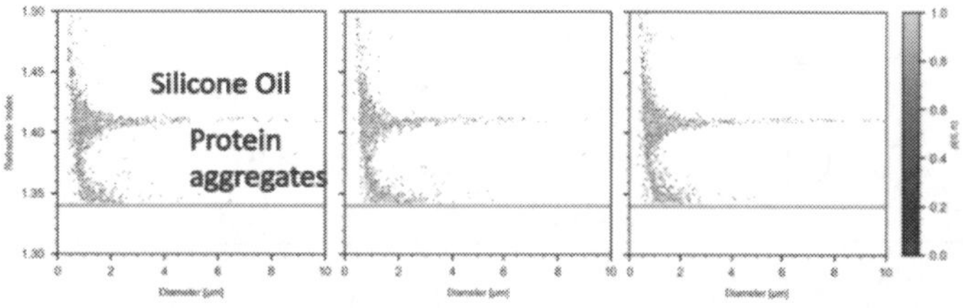


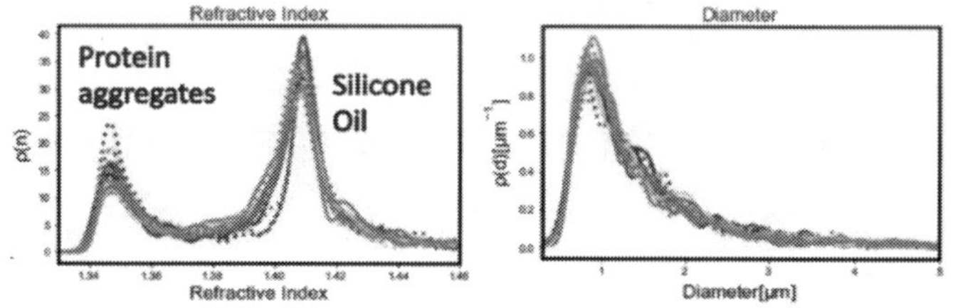


- 16 independent measurements of the same sample of IgG + Silicone Oil
- Concentration Results for 16 measurements:
  IgG = $2.9 \pm 0.3 \times 10^5$ particles/mL
  Oil = $6.6 \pm 1.0 \times 10^5$ particles/mL
  Total = $1.0 \pm 0.1 \times 10^6$ particles/mL

5

# User Friendly

- Easy sample loading
- xCell disposable chip ensures:
  - No cross-contamination
  - No cleaning

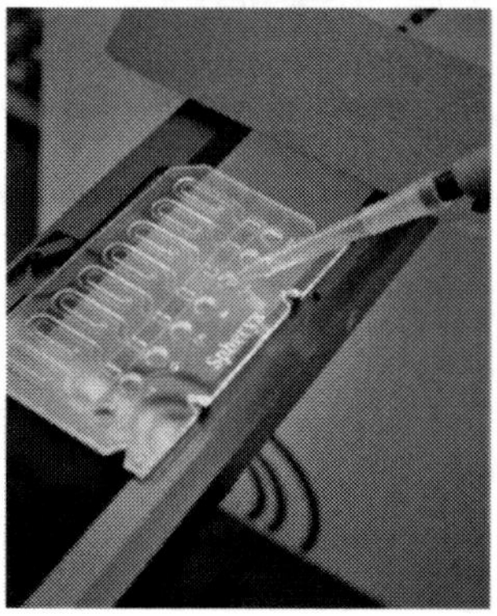

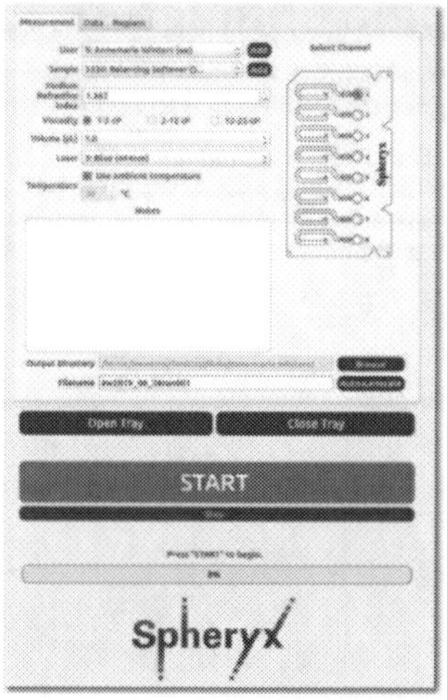


- Load xCell & Press Start
- Typical measurement 10 min

- xStream Autosampler

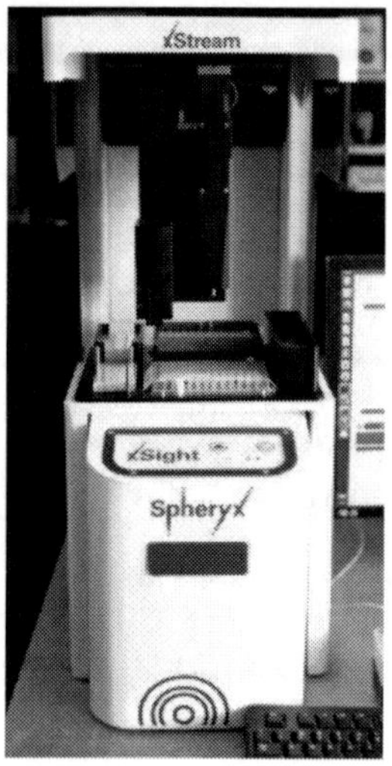


6

# Validation: Scanning Electron Microscopy

SEM image: 500 nm
polystyrene spheres

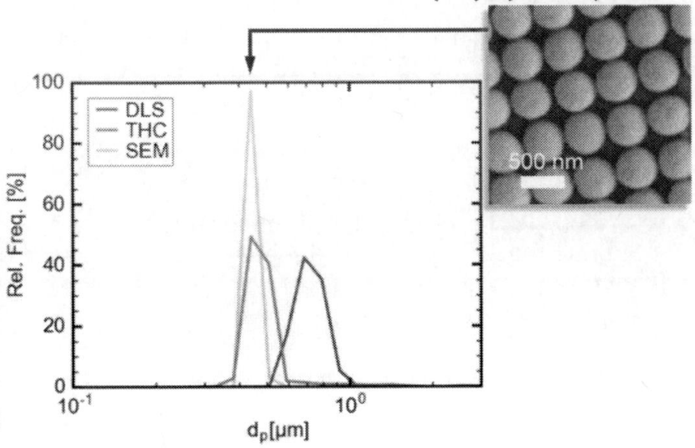


S

THC Benefits:
- In situ
- Fast
- Minimal sample prep
- Minimal training

7

# Dilution Study

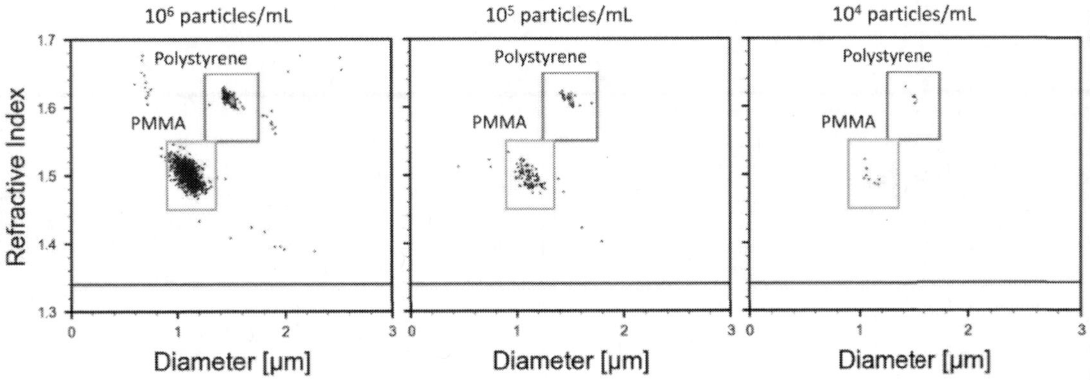


- Varying the concentration of 1.5 polystyrene and 1.14 PMMA beads in particles/mL over multiple orders of magnitude

8

## Industry-Leading Concentration Accuracy

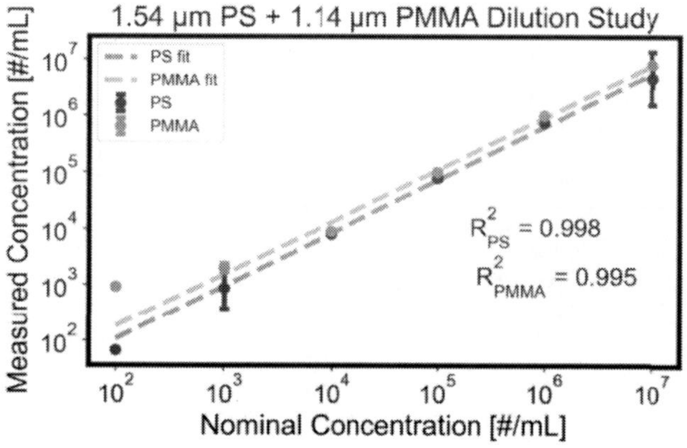


- Several orders of magnitude of dilution with accurate concentration measurement
- Concentration, size and index of different particles in a heterogeneous mixture
- Limit of high concentration is particle overlap
- Limit of low concentration is sampling statistics

9

## Air Bubbles + Silicone Oil in Gelatin and Sucrose

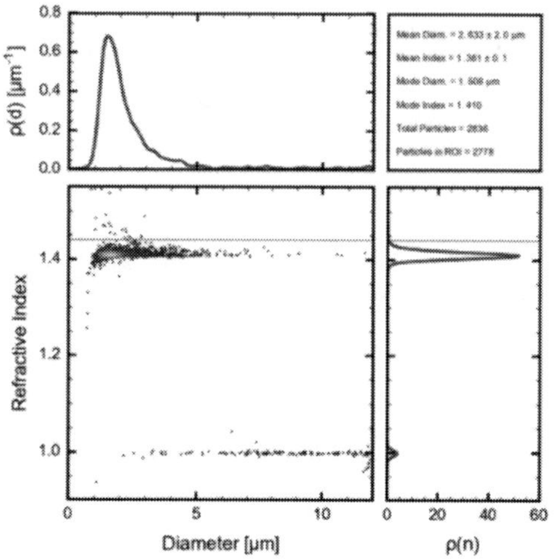


Air bubbles are easily distinguished from other contaminants

Mixture concentrations:
Silicone Oil +
1.5625mg/mL gelatin +
1.51M Sucrose

# Measuring Multiple Oils Simultaneously

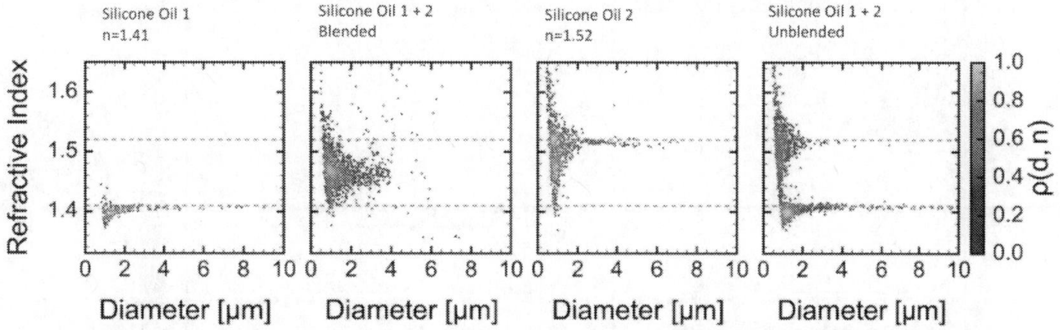


- 2 Silicone oils of different index (n(oil 1)=1.52 and n(oil 2)=1.41) are mixed in 2 different ways
- The samples are prepared: 1) Emulsify and then mix the emulsions OR 2) Blend the oils and then emulsify
- If the oils are not blended, they do not mix in the emulsion and two separate THC emulsion signatures appear
- If the oils are blended 1:1 the THC has a single emulsion signature at the average index of the two oils (n=(n1+n2)/2)

11

# Measuring Multiple Oils: Density Plots

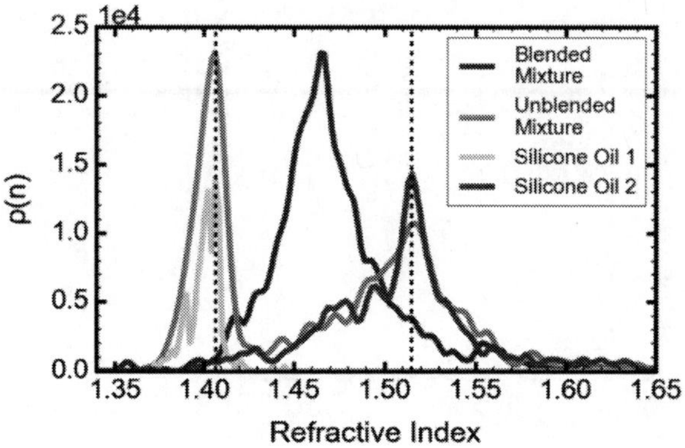


- Different silicone oils are easily distinguished by the refractive index
- Blended oils (1:1) have the refractive index equal to the average of the refractive indexes of the 2 oils
- If the oils are not blended, they do not mix in the emulsion and two separate THC emulsion signatures appear

12

# Mixtures of Oils of Different Refractive Index

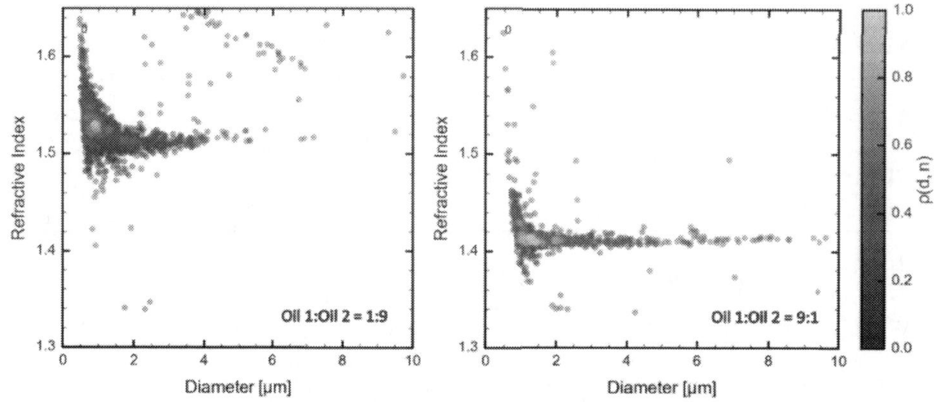


- The ratio of the amounts of each oil determines the refractive index of the mixture
- The mixture that is 10% Oil 1 and 90% Oil 2: n = 0.1 n(oil 1) + 0.9 n (oil 2)
- The mixture that is 90% Oil 1 and 10% Oil 2: n = 0.9 n(oil 1) + 0.1 n (oil 2)
- n(oil 1) = 1.41 and n(oil 2) = 1.52

13

# Density Plot of Different Mixtures of Oils

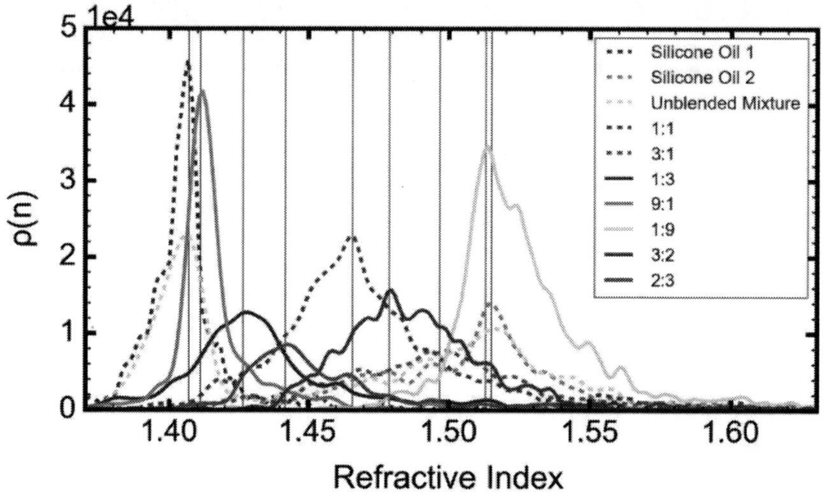


- The refractive index of the blended mixtures depend on the ratio of the two oils in the mixture and can be predicted by the weighted average of the indexes of the two oils in the mixture

14

## Refractive Index of Different Mixtures of Oils

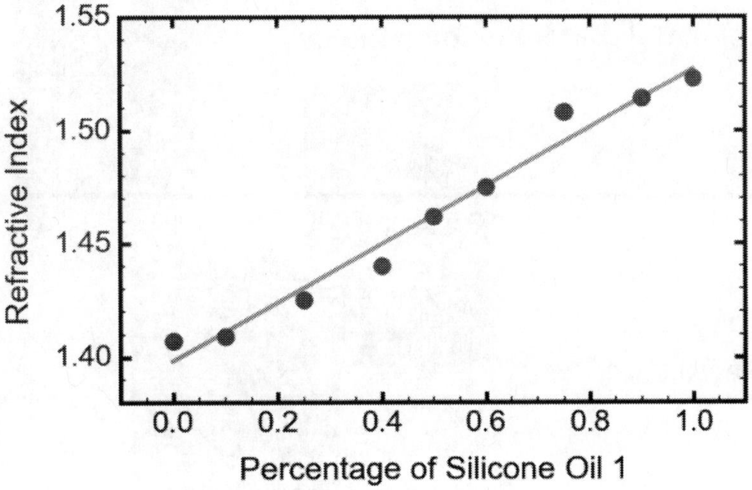


- The refractive index of the blended mixtures depend on the ratio of the two oils in the mixture and can be predicted by the weighted average of the indexes of the two oils in the mixture

15

## THC Measurements of Ingredients Found in Cosmetics

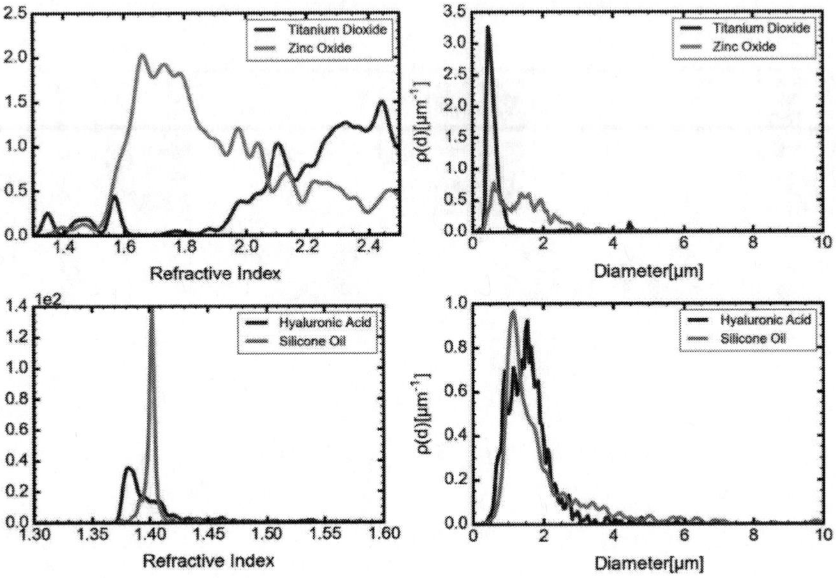


- Size alone is not sufficient when there are multiple ingredients present in the sample
- Index of refraction provides an additional dimension of information to quantitatively identify multiple components - even when they are the same size

16

## Real World Cosmetic Samples

- Can we distinguish Different Components using THC?
- Samples:
- L'Oreal Shampoos
- L'Oreal Moisturizers
- Shiseido Facial Care

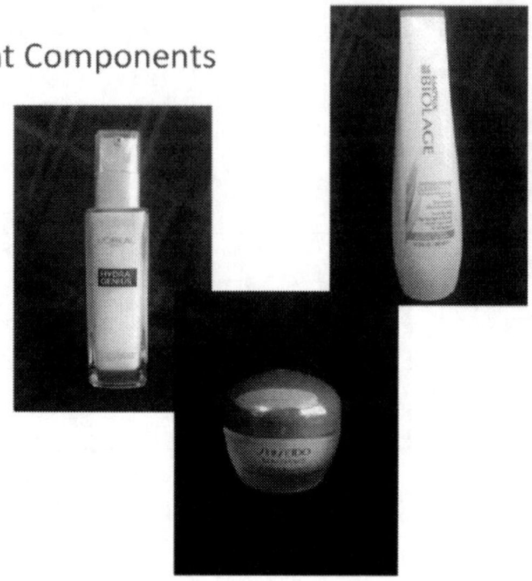


Spheryx

17

## THC of L'Oreal Shampoos

Lemongrass　　Aloe　　Rosemary

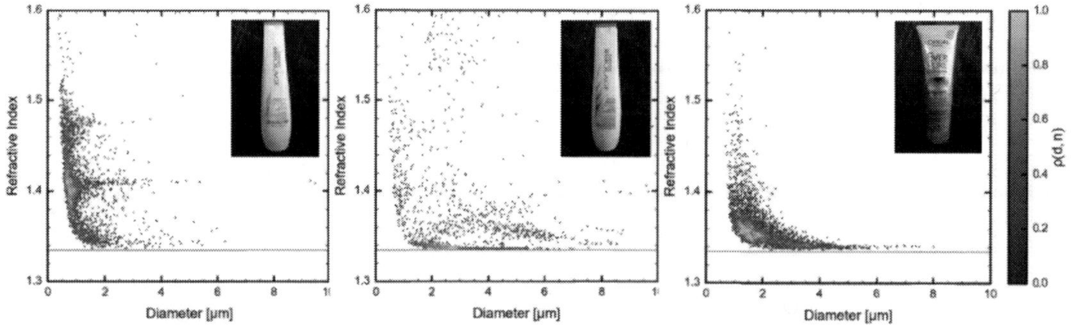


Each shampoo has its own unique THC signature

18

# THC of L'Oreal Shampoos

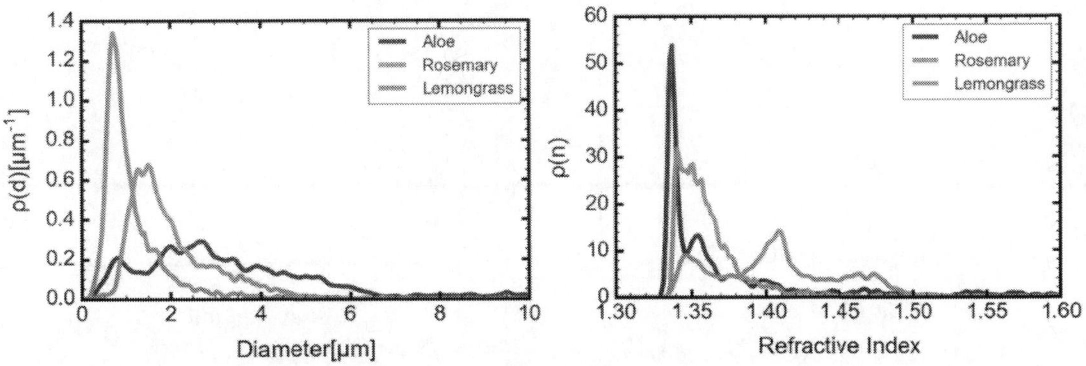


- Refractive index provides an added dimension of information to identify different components in the sample.
- Note the distinct signature of higher index emulsions in the lemongrass sample at n=1.41 and n=1.47 as well as particles at lower index. The size distribution alone does not distinguish between the different components.

19

# THC of L'Oreal Moisturizers: L'Oreal Paris & L'Oreal CeraVe

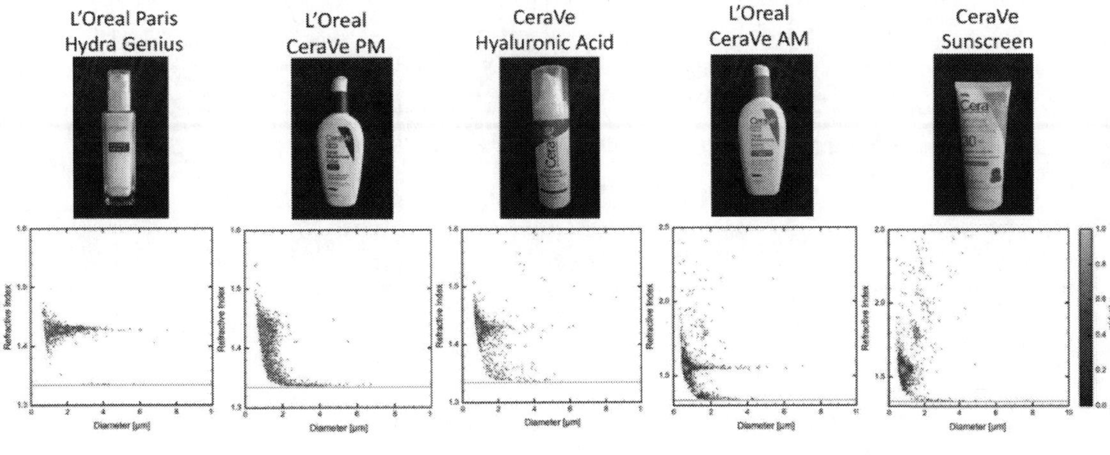


Each moisturizer has its own unique THC signature
(Note: Different refractive index scales)

20

# THC of L'Oreal Moisturizers

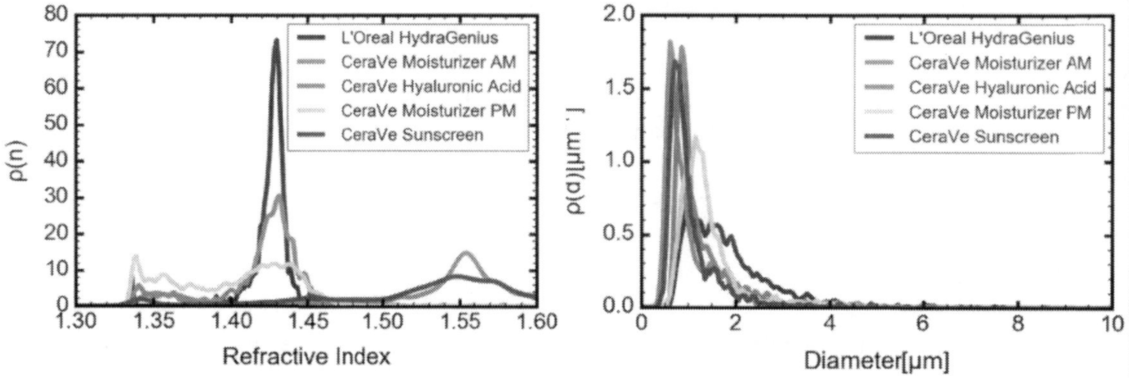


- Refractive index provides an added dimension of information to identify different components in the sample.
- All of the samples have signatures of emulsions, but at different refractive indexes. Different indexes identify different ingredients.
- CeraVe Sunscreen and CeraVe AM, both have a similar THC signature. Note the high index species characteristic of ingredients used for sun protection.

21

# THC of Shisheido Facial Products

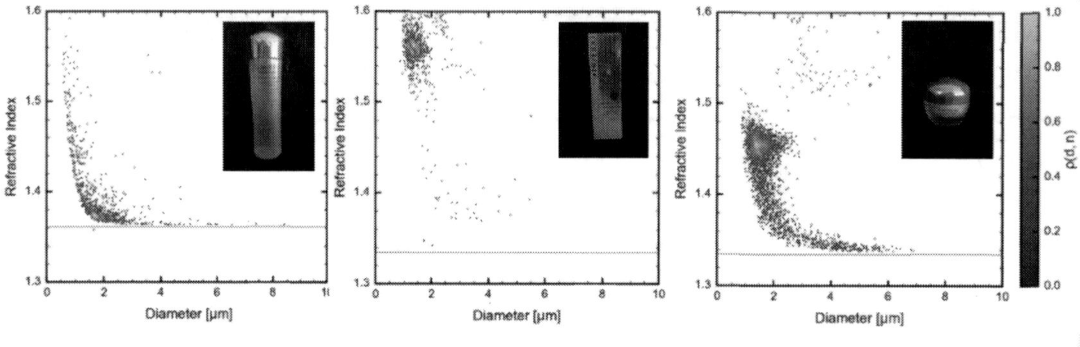


Each product has its own unique THC signature

22

# THC of Shisheido Facial Products

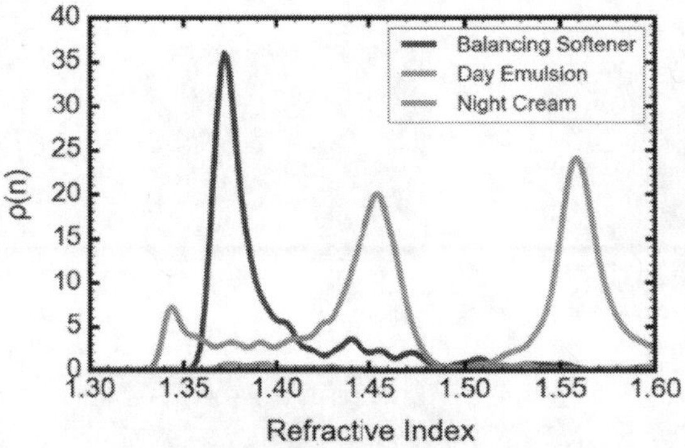


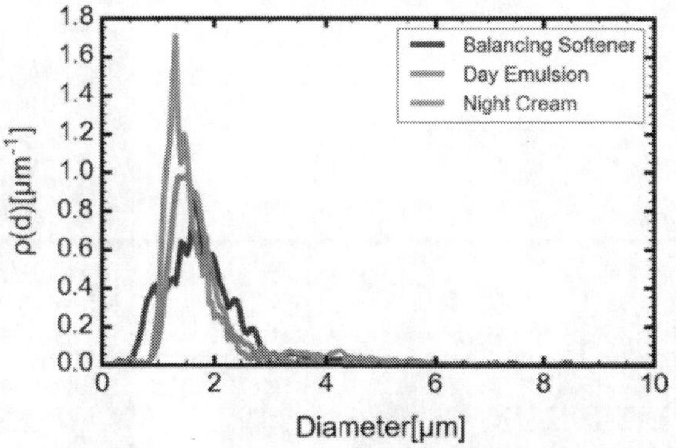


- The size distributions are not sufficient to identify the different products. The refractive index distributions of the three products are unique and distinguishable.
- The Day Emulsion has a high index species consistent with ingredients used for sun protection.

23

# Conclusions

- THC detects multiple different components in complex cosmetic products
  - Identifies components
  - Relative concentrations
- Applications
  - Quality control for ingredients
  - Aging studies for shelf-life monitoring
  - Formulation development
  - Product comparisons

Spheryx

24

Spheryx Products
PRISM AWARDS WINNER
R&D 100
xSight
xStream
xCell
Spheryx

# Polysaccharides for Prevention of Epidermal Microbial Transfer

## Authors

Fazilova, S. - Presenter, Active Concepts LLC
Drohan, M.

In the last year, masks have served as physical barriers to prevent transfer of aerosolized pathogens. But what about contact transfer? Reaching for door handles, greeting others with a handshake, or signing documents with a shared pen were once tasks performed subconsciously. It is now apparent how frequently microorganisms, pollution, and dirt transfer from surrounding surfaces to hands and skin. Hand sanitizers used to kill surface microorganisms are often associated with skin irritation and dryness. The skin microbiome is an ecosystem of microbial communities that inhabit the skin to maintain homeostasis. The non-pathogenic bacteria that normally reside on the skin, or commensal bacteria, becomes imbalanced after the use of hand sanitizers. The indiscriminate disruption of the skin microbiome caused by cidal actives in hand sanitizers can destroy commensal bacteria and make the skin more vulnerable to incursion by pathogenic organisms.

Polysaccharides and biopolymers are known to provide barrier protection. Previous research investigating classes of natural materials to prevent transfer of bacteria lead to the nanopattern surface of Clanger cicada (*Psaltoda claripennis*). The wings of this insect serve as a model for the development of innovative functional surfaces that offer increased resistance to microbial contamination. This piqued the interest in examining the effects of a cosmetic active rich in polysaccharides to prevent the transfer of bacteria.

Interestingly, many succulent plants rely on polysaccharides to maintain cellular turgor. The Resurrection Plant provides a compelling case as the extracted polysaccharides and biopolymers prevent microbial transfer, aid in pollution protection, and increase wound healing capabilities. In this study, we discuss the potential use of a naturally derived polysaccharide rich extract, *Selaginella Lepidophylla* Extract, as a novel defensive shield against microbes compared to other known polysaccharides.

# Trends in Product Development

**Lucy Gildea, Ph.D.**
Chief Innovation Officer, Product & Science
Mary Kay Inc.

MARY KAY 1

## Outline

- Intro - Product Development
- General Trends Affecting "How"
- Specific Trends in Methods and Tools
  - Consumer Research
  - Package Engineering
  - Product Formulation
  - Analytical / Stability
  - Process Development
- Conclusion

MARY KAY 2

# Consumer Product Development

Focus of this keynote is on –

- **How** product development is done

  For example, affected by trends like digitalization and digital tools

  *and **not** on*

- **What** products are developed

  For example, affected by trends like clean beauty, personalization, sustainability

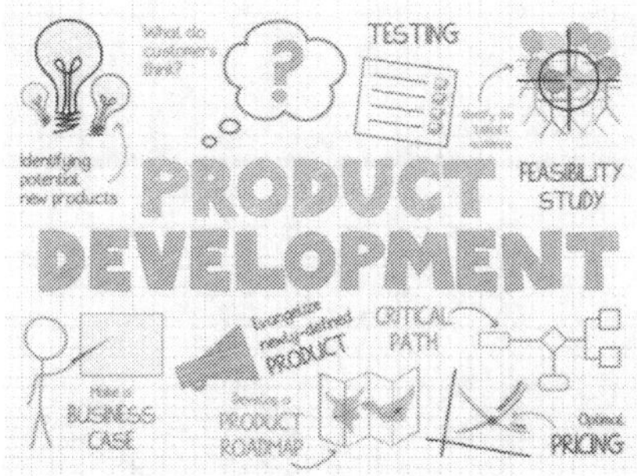




# Intro - Product Development Approach

*Overall, product development approach is largely common across companies doing cosmetic and other consumer product development*

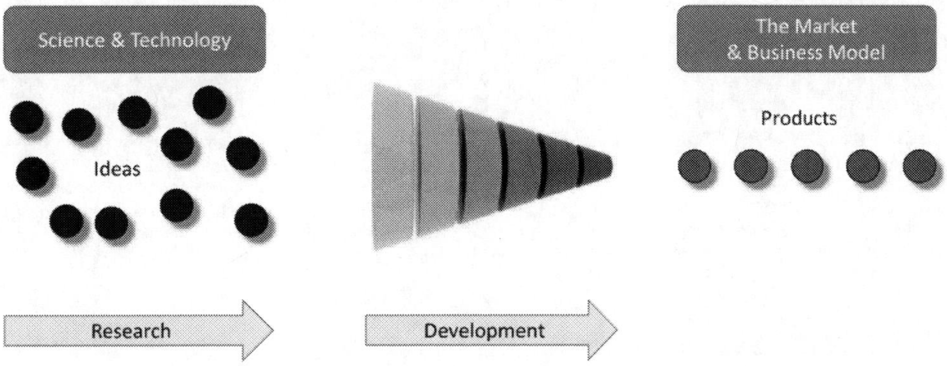

## Transformation of Product Development Process

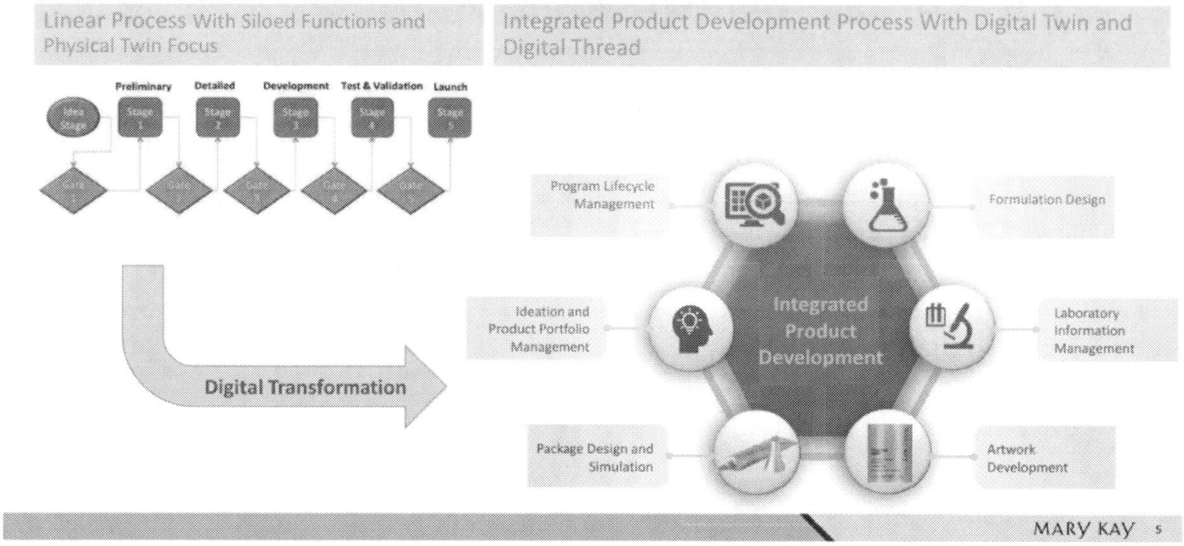


## Product Development – Innovative, Faster, More Efficient

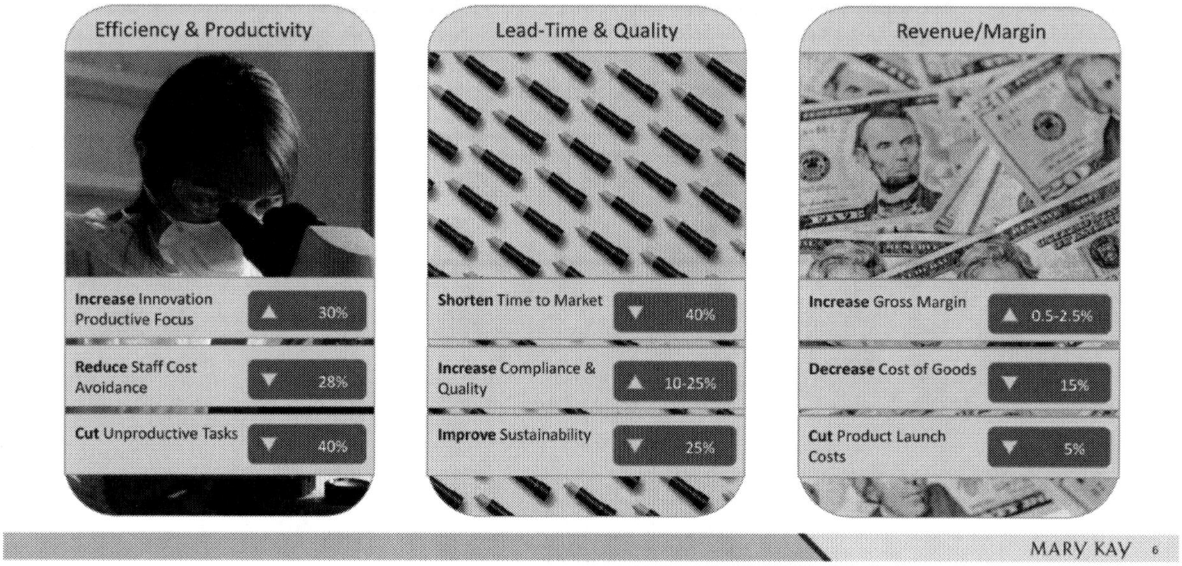

# General Trends

*Some of the biggest trends affecting how product development is done originate in the digital revolution and software technology advances*

- Mass adoption of 3D printing
  - Printing of prototypes, particularly packaging
- Social media, and rapid product development feedback
  - Iterative evaluation of product using digital platforms
- Digital design lowering costs and increasing speed
  - Digitally created products, packages, and product experiences
- Stability prediction based on AI/ML examination of large stability/formulation databases
- Physical Modeling - Computational Fluid Dynamics for process development
- Multi-scale simulation for formulated products (chemistry and physical chemistry modeling)
- Product performance prediction based on AI/ML of large consumer/performance databases

MARY KAY 7

# Trends – Consumer Product Research

*Identifying Consumer Needs*

- Consumer Sampling Strategy
  - Consumer feedback is dependent on the parameters in which participants are selected
    - Common factors include awareness or usage of a product, age, purchasing habits, etc.
  - Traditional demographics are changing with a growing multiethnic population, evolving gender identification and expression, and an increased focus on inclusivity
  - Benefits:
    - Previously underserved groups now have their product needs met with inclusive designing during product development
  - Watchouts:
    - Cannibalization of top selling products with many targeted design products can reduce SKU productivity- striking a balance is key!

MARY KAY 8

# Trends – Consumer Product Research

*Identifying Consumer Needs*

- Social Listening & AI Data Analysis
  - Mining data from user generated content can eliminate the need for primary research while still gaining actionable insights
  - Provides access to previously "hard to reach" consumer groups
  - Benefits:
    - Social listening tools are generally more cost effective than multiple primary research studies
    - Learning/using consumer language vs. internal terminology
  - Watchouts:
    - Many people don't participate in product discussion/feedback online so some groups could be underrepresented
    - Distinguishing between a trend and a fad can be difficult without monitoring over time



# Trends – Consumer Product Research

*Technology Advancements*

- Mobile Surveys Design
  - Survey durations must be shorter and more engaging! People are taking these surveys on the go
  - Survey questions and content need to have easy user experience on touch screens or smaller screen sizes
  - Benefits:
    - Can intercept users in various stages of product engagement
    - Can collect feedback or reviews from those who purchase the products in real time
  - Watchouts:
    - Attention spans are limited so fighting through the noise of other media can reduce response quality
    - Not all research methodologies are best suited to a mobile survey experience

# Trends – Consumer Product Research

*Technology Advancements*

- Virtual Prototyping
  - 2D & 3D rendering or videos of product prototypes before a product is ever created in real life
  - Consumers can even co-create products by modifying/selecting key benefits or features
  - Benefits:
    - Gain consumer buy in on the idea before investing resources: fail fast, fail cheap
    - Evaluate numerous ideas before narrowing down to final product selection
  - Watchouts:
    - Some attributes can't be replicated online (scent, how a product feels, etc.)
    - When sharing new forms/benefits, additional education might be required before consumers can provide actionable feedback

MARY KAY 11

# Trends – Lab of the Future

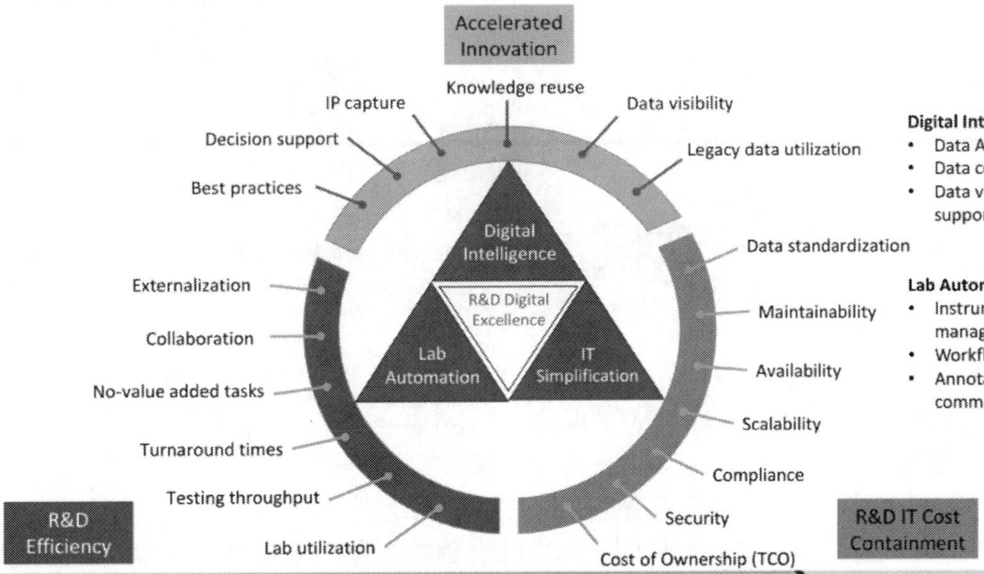


**Digital Intelligence**
- Data Access
- Data contextualization
- Data visualization and decisio support

**Lab Automation**
- Instrument integration & data management
- Workflow support
- Annotation and communication

MARY KAY 12

# Trends – Product Formulation

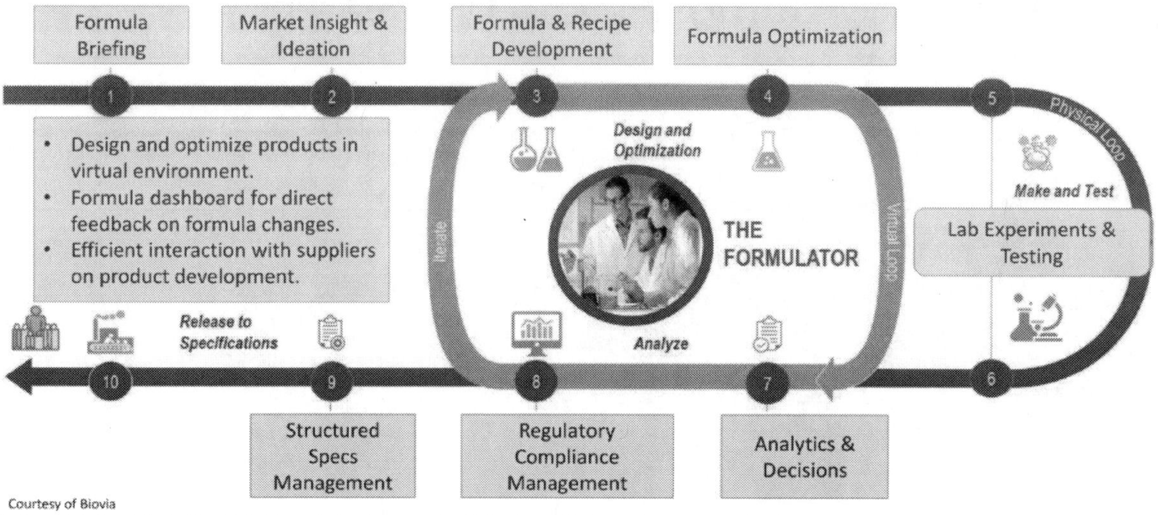


Courtesy of Biovia



# Trends – Multi-Scale Modeling

*Simulation from Chemical to Product Scale*

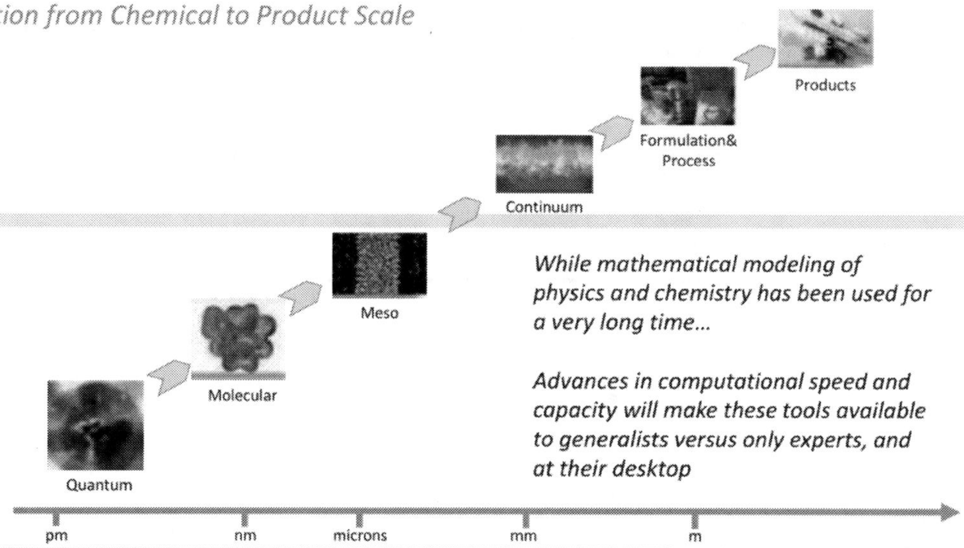


*While mathematical modeling of physics and chemistry has been used for a very long time...*

*Advances in computational speed and capacity will make these tools available to generalists versus only experts, and at their desktop*

# Trends - Data Science

*Artificial Intelligence / Machine Learning*

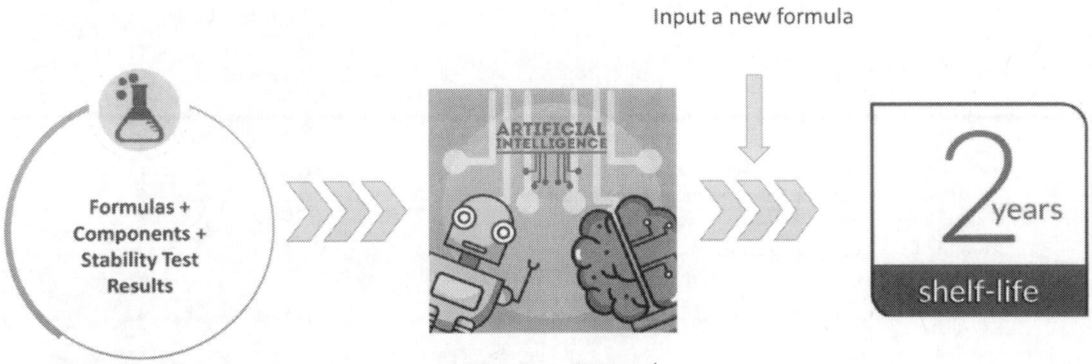


**All historical data** on formulas, properties, and stability results

Artificial Intelligence/ Machine Learning engine creates a model

Output a predicted stability profile



# Trends – Package Engineering

*Package Design, Copy/Claims, and Artwork all in one System*

1. Link ideation to industrial design with integrated workflow
2. Integrate design and simulation
3. Improve testing by conducting performance simulations
4. Capture and reuse company standards
5. Manage all elements of a package in a single repository for managing and approving copy assets
6. Maintain control from copy creation to final artwork approval
7. Use simulation to determine how a new package design will behave at every stage of its lifecycle
8. Create photorealistic images to accelerate package validation without costly prototypes

# Trends – Process Development

*Computational tools and simulation*

- Computational tools for simulating and modeling physical and chemical systems have been around for a considerable time
  - However, the ease, power and speed of computation has been growing exponentially to the point where desktop computers can handle very complex systems, enabling -
    - For example, the use Computational Fluid Dynamics (CFD) to simulate key formulated product attributes, like emulsion stability, particle size distribution and viscosity, etc. in order to optimize processing parameters.
    - And reduce the number of costly pilot and production trials with increased success in batch scale-up.

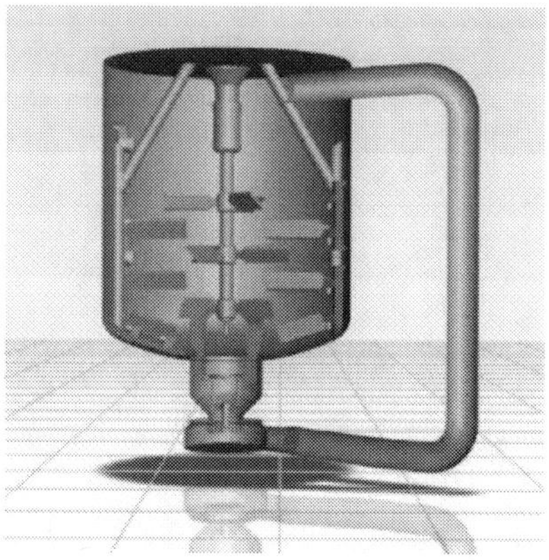

MARY KAY 17

# Trends – The Final Frontier?

*Artificial Intelligence will transform product development*

**Search** → **Synthesize**

***Query: Find me all information for search terms "sunscreen" and "SPF 15"***

***Query: Formulate for me a global SPF 15 sunscreen that maximizes consumer perceived skin conditioning and minimizes cost with a robust making process***

MARY KAY 18

# Conclusion

*Trends in Product Development – Digital and Data Revolution*

- Digital development platforms and data acquisition and analysis will change the way we develop and evaluate products
- Advances in digitalization and digital tools will continue to speed and reduce the cost of the product development
  - Single, end-to-end Product Lifecycle Management (PLM) platform with forward and backward continuous "data threads"
  - Models for virtual product creation and testing and prediction
- The development of more sophisticated tools, like those involving artificial intelligence, will only accelerate the process

MARY KAY 19

# Questions?

*Trends in Product Development – Digital and Data Revolution*

MARY KAY 20

# AUTHOR INDEX